S0-DSG-338

The Sun Beneath the Sea

"BEN FRANKLIN"
1969
GULFSTREAM DRIFT MISSION

THE SUN BENEATH THE SEA

Jacques Piccard

TRANSLATED FROM THE FRENCH BY DENVER LINDLEY

NEW YORK CHARLES SCRIBNER'S SONS

Picture credits

Erwin Aebersold:
pages xxviii (top), xxix, xxxii (top), xxxiii, xxxiv (bottom), xxxvi (bottom).
Grumman Aerospace Corporation:
pages xxii–xxiii, xxv (top), xxvi, xxvii (top and bottom), xxx, xxxviii (top and bottom), xxxix.
NASA:
pages xxxiv (top), xxxv (top and bottom).
Photo Pfister:
pages xviii–xix, xx, xxi.

All photographs not otherwise credited are by the author.

A—1.70 [H]

Printed in the United States of America
SBN 684-31101-1
Library of Congress Catalog Card Number 76-123854

This book is dedicated to Grumman, an aerospace corporation which realized that the sea is as vital to the future of mankind as space itself and which contributed simultaneously to the moon exploration with the Lunar Module and to the Gulf Stream exploration with the *Ben Franklin*

CONTENTS

PART TWO: MESOSCAPH *BEN FRANKLIN*

PART THREE: THE LONGEST NIGHT

APPENDIX: RESULTS OF THE GULF STREAM DRIFT MISSION

FOREWORD

While a fascinated world watched the Apollo 11 moon expedition in July 1969, another very important scientific voyage was taking place. With similar goals of exploring unknown places, seeking answers to previously unanswered questions, and recording scientific data, Jacques Piccard was leading an undersea drift mission in the Gulf Stream off the East Coast of North America.

Both the immensity of the oceans of earth and the vastness of the space surrounding earth have captured the curiosity of men since the dawning of knowledge. One can, from the surface, peer down into the depths of the ocean and also upward at the heavens. But fully to explore either of these frontiers requires extensive new technology. And, in some ways, the technology requirements are startlingly similar.

Dr. Piccard designed the vessel, the *Ben Franklin,* which, when submerged, must provide a livable environment for its crew, as a spacecraft does outside the earth's atmosphere. During long voyages, the crew must live and work together in a situation such as might be encountered during long stays in an earth-orbiting space station.

Because of these similarities, a young NASA engineer, Chester B. May, was allowed to accompany the drift mission to observe and evaluate the crew at work, rest, and play and to relate the experience to the development of future NASA space stations. NASA is also concerned with the Tektite Program, in which scientists spend long periods in habitats on the ocean floor. Several NASA employees are involved in that operation, studying habitability and observing the crews from both a physiological and a psychological standpoint.

You will enjoy Dr. Piccard's story, with the mystery of the ocean, the excitement of the mission itself, and the scientific findings. His achievement is also a portent of the future, when more and more expeditions will explore the secrets of the vast ocean depths which are filled with life-sustaining food and mineral riches to enhance life on earth for mankind.

I think there is a common bond—or a common dream, if you wish—between the men who explore these two vast unknowns—the oceans and space. It is a fervent hope that nationalistic aggressions will cease and fade into a new age of knowledge and cooperation, that the very immensity of the ocean depths and the uncharted avenues of space will show that, to extract their benefits, the nations of man must work as one.

As you travel with Dr. Piccard you will be in the company of men from several different nations. They are in search of a brighter future for all mankind. Their language is a common expression of hope, and their message is one that today's world needs.

WERNHER VON BRAUN
Deputy Associate Administrator,
National Aeronautics and Space Administration

PREFACE

In the seas, as on the land, all life, all movement, stems ultimately from solar energy, transformed sometimes over millions of years, sometimes in a few days or even a few hours. The phytoplankton, those vegetable algae that teem in the sea and are the beginning of the marine food chain, through photosynthesis absorb the light of the sun and are sustained by it; a similar phenomenon occurs with vegetation on land. All natural fuels—wood, coal, oil—in burning give back the solar energy acquired at the time of their formation. Hydraulic energy, produced by natural waterfalls and man-made dams, is made possible because water was evaporated by the heat of the sun, to fall later as rain, snow, or sleet and accumulate to supply the power plants that produce electricity. The various forces of nature—winds, waves, storms, ocean currents—derive from the sun and would disappear if the sun were extinguished.

Not only is the sun the source of swarming life that the *Ben Franklin*'s searchlights revealed beneath the waves; from the sun also came the power that charged the searchlight batteries. During thirty days and thirty nights the sun accompanied us everywhere; in the obscurity of the deeps the sun beneath the sea kept its rendezvous.

In the *Ben Franklin*'s construction and its historic voyage, so many people made notable contributions that it is impossible to make full acknowledgment here. But I would like to express my warmest gratitude to a number of organizations: Giovanola, Monthey, Switzerland, which built the hulls of the two mesoscaphs; Electrona, Boudry, Switzerland, which developed with us the new battery system used on board; Brown, Boveri and Co., Baden, Switzerland, and AEG and Pleuger, Hamburg, Germany, which supplied the electrical systems and propulsion motors; and finally Danzas, Geneva, Switzerland, which made possible the transportation of the *Ben Franklin* from Swiss mountains to Palm Beach, Florida, the final home port of the mesoscaph. Apparent throughout this book is the close cooperation provided by Grumman Aerospace Corporation, NASA, and the U.S. Navy—mainly the Naval Oceanographic Office. Without that perfect understanding, the Gulf Stream Drift Mission would never have been realized. To all who contributed to the preparatory work and the final mission, my heartiest thanks.

JACQUES PICCARD

A PORTFOLIO OF PICTURES

The mesoscaph *Auguste Piccard*

Interior *(above)* and cockpit *(right)* of the *Auguste Piccard*

HEAVY LIFT
SERVICES INC.
PORT OF PALM BEACH
BEN FRANKLI
WARNING
POWER PLANT INTAKE
ALL VESSELS KEEP STERN CLEAR
OF THIS AREA — ALWAYS DIRECT
STERN TO THE NORTH OR EAST

The mesoscaph *Ben Franklin*

(Right) Construction of the *Ben Franklin: (above)* rolling the main plates; *(below)* the two parts of the hull

The Giovanola factory at Monthey, Switzerland, where the *Auguste Piccard* and the *Ben Franklin* were built

The *Ben Franklin: (above)* in drydock at Palm Beach, Florida; *(right, top)* at sea; *(right, bottom)* under water during a test dive off Palm Beach

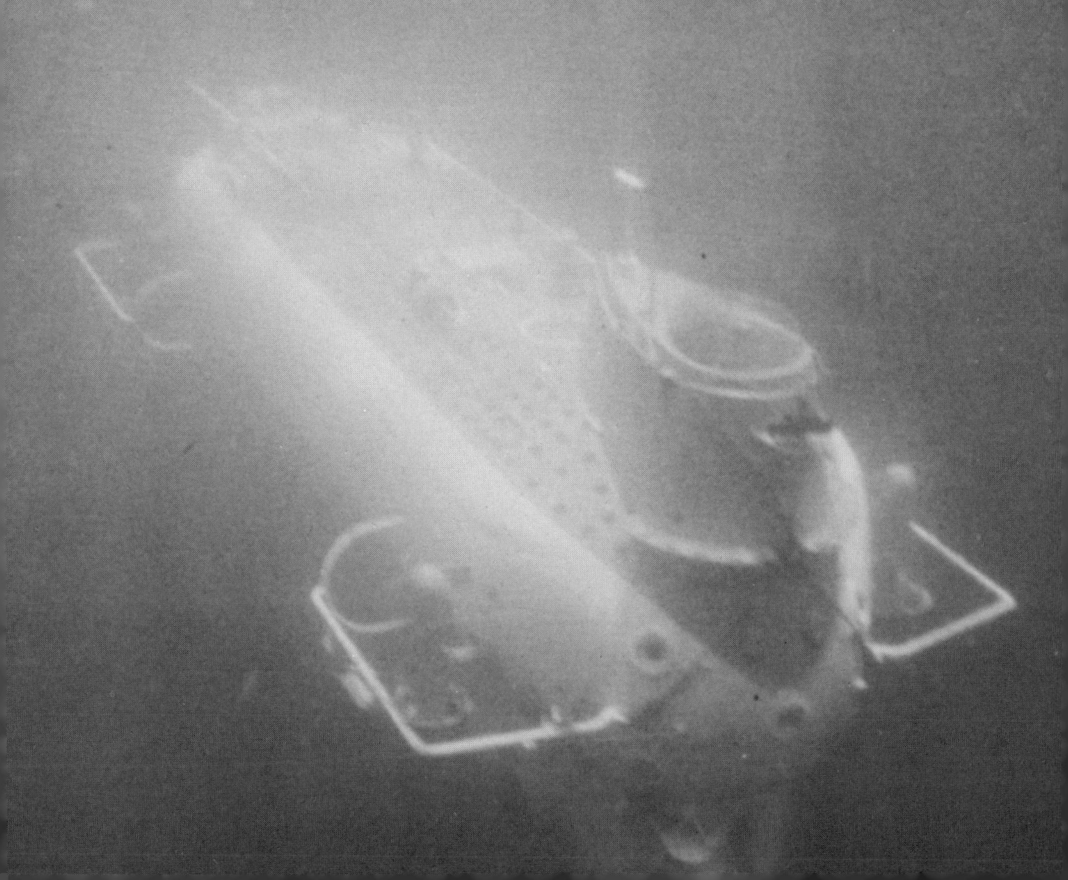

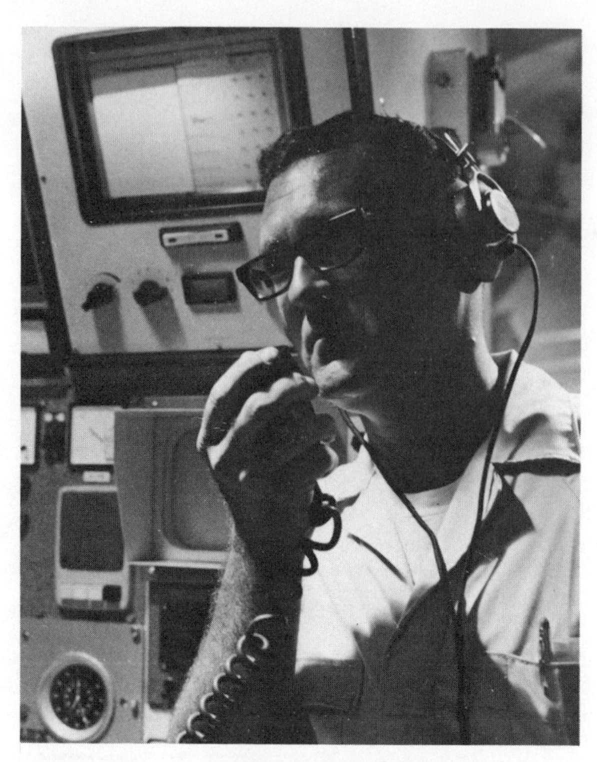

Don Kazimir of Grumman, captain

Erwin Aebersold, pilot, in the cockpit, facing the Rolex timing equipment and the navigation instruments, and holding the motor controls in his hands

Jacques Piccard, leader of the mission

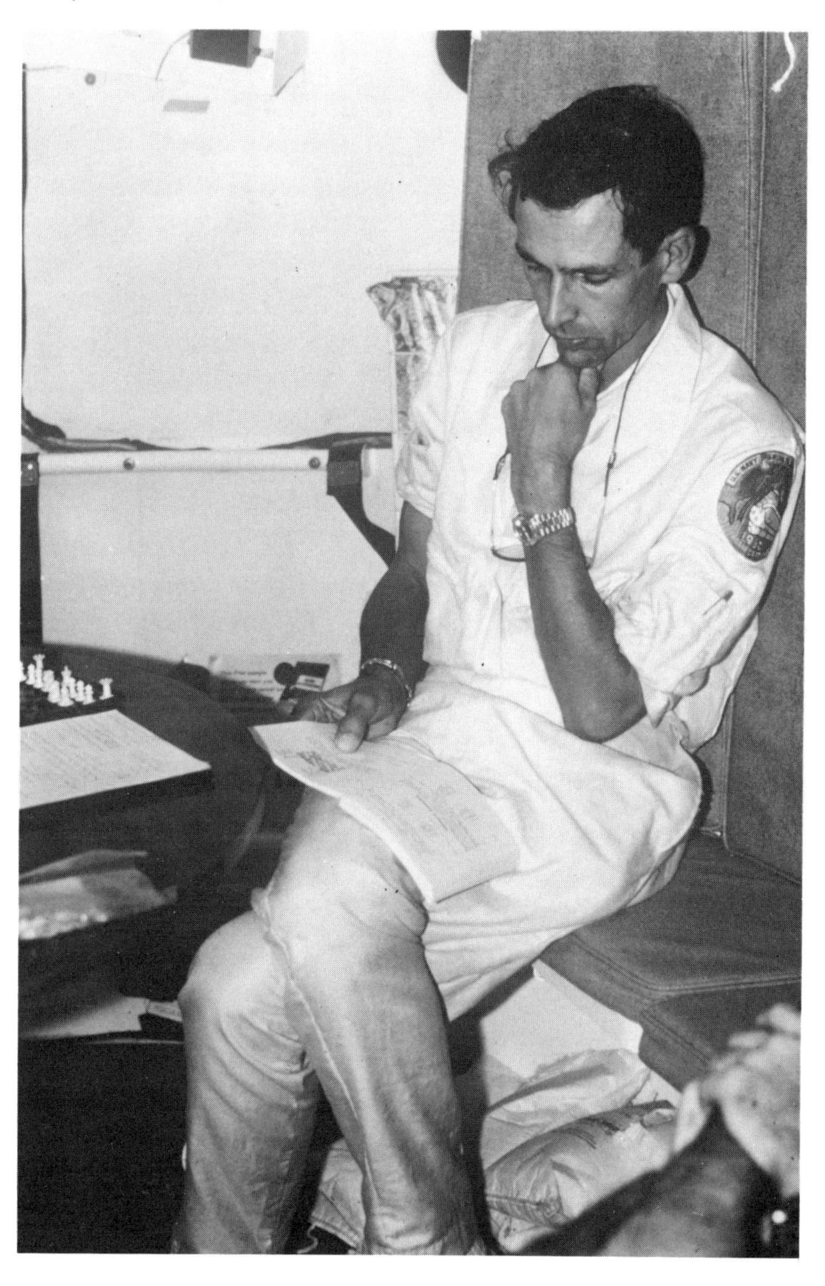

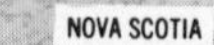

BOSTON

CAPE COD

NEW YORK CITY

14 AUG

BALTIMORE

CAPE HATTERAS

ATLANTIC OCEAN

CHARLESTON

WEST PALM BEACH

14 JULY

GRUMMAN / PICCARD
GULF STREAM DRIFT MISSION

ROUTE OF THE *BEN FRANKLIN* SUBMERSIBLE

14 JULY – 14 AUGUST 1969
DISTANCE: – 1444 NAUTICAL MILES

——— BEN BRANKLIN
- - - - - GULF STREAM

HISPANIOLA

A shoal of tuna at a depth
of 200 meters, photographed by natural light

(Left) The route of the mission

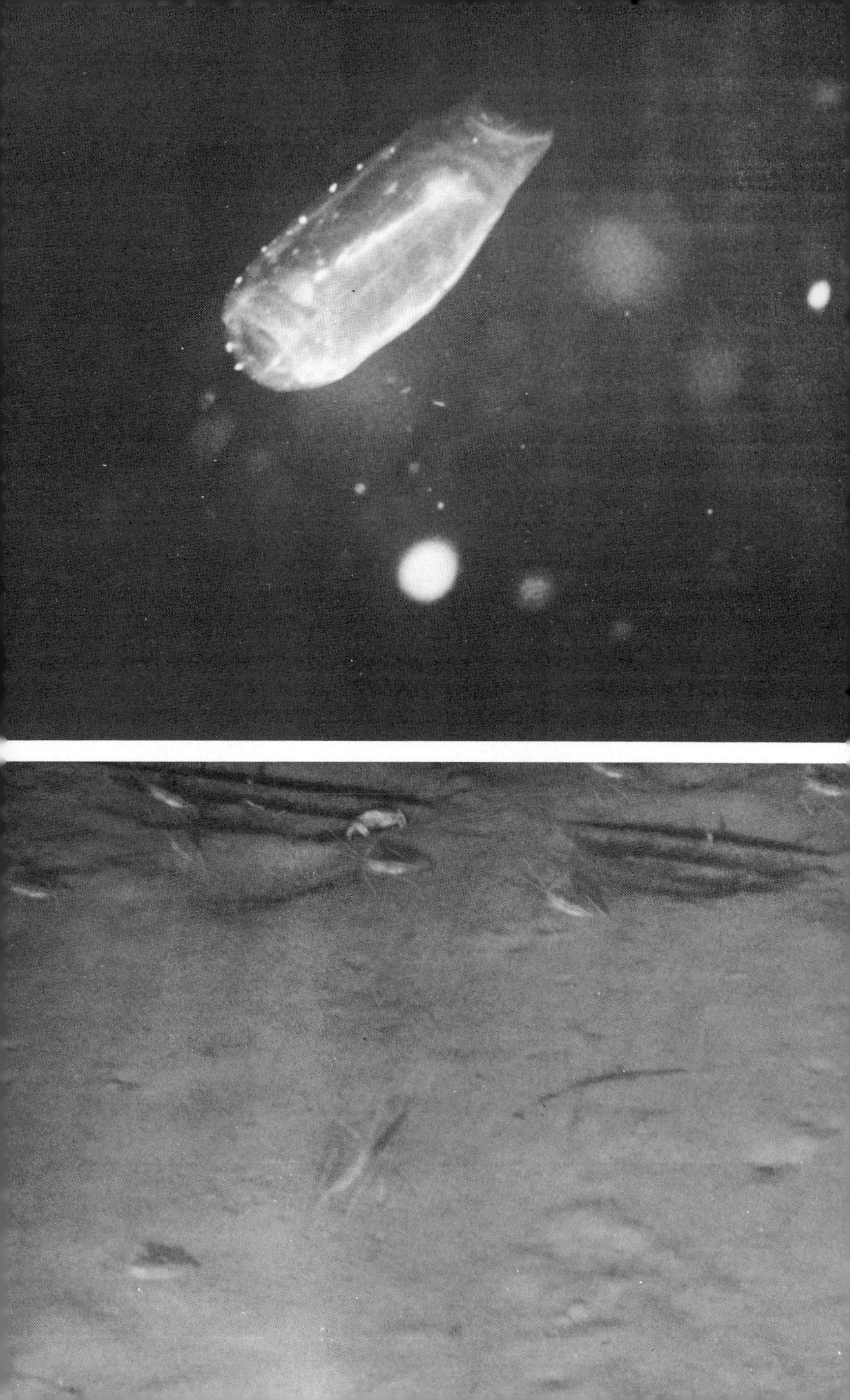

On the sea bottom, a crab about 30 centimeters wide

(Left, top) A salpa,
about 2.5 centimeters long, one of
many observed during the drift
(Left, bottom) Crustaceans surrounding
the telephone cable connecting
Florida and the Bahamas

Frank Busby, of the Navy Oceanographic Office, at work

Chet May, NASA engineer and observer, resting in a helmet designed by NASA for taking electroencephalograms at night

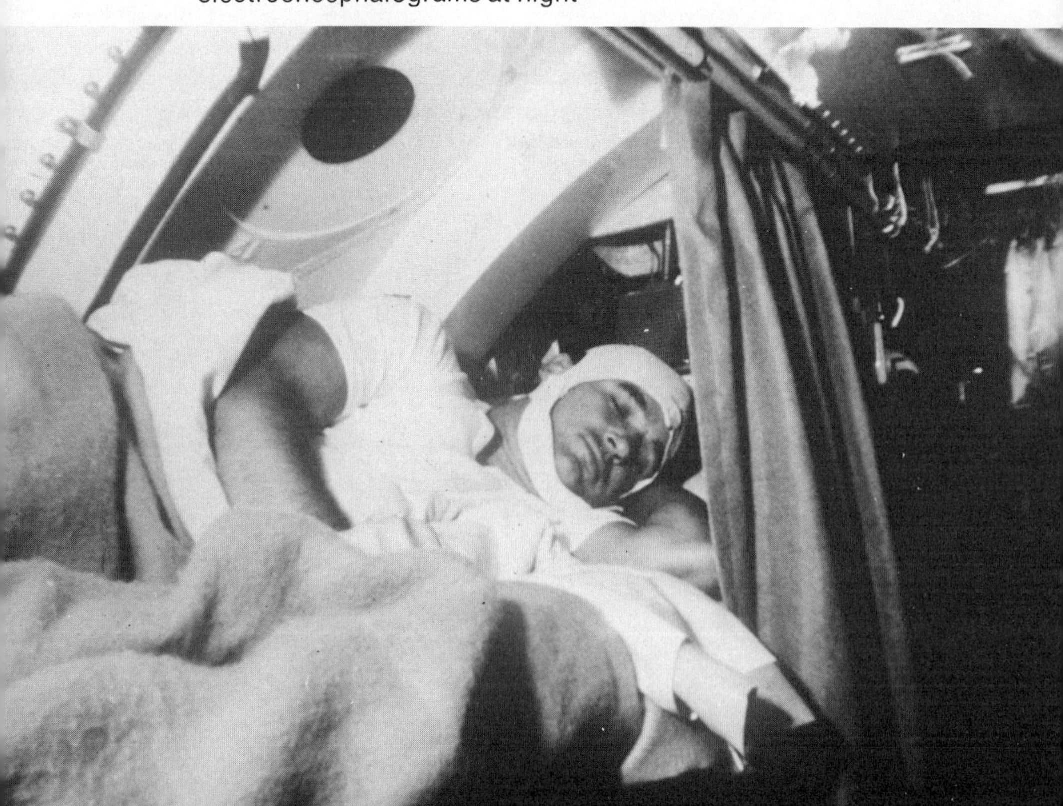

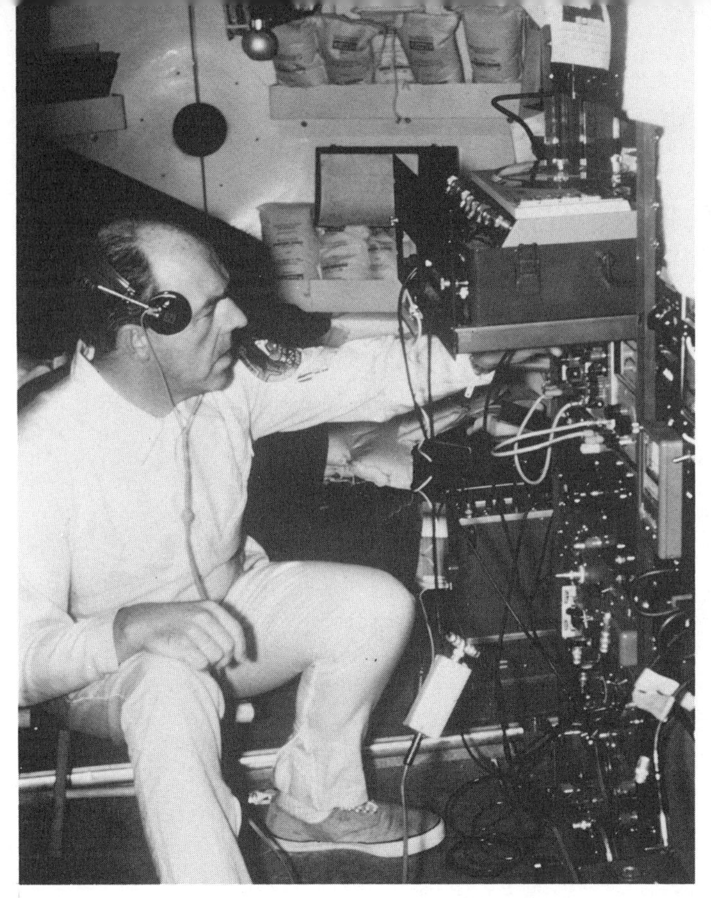

Ken Haigh, oceanographer and acoustic specialist,
with part of his equipment

Aebersold and Piccard commemorating
the Swiss national holiday, August 1

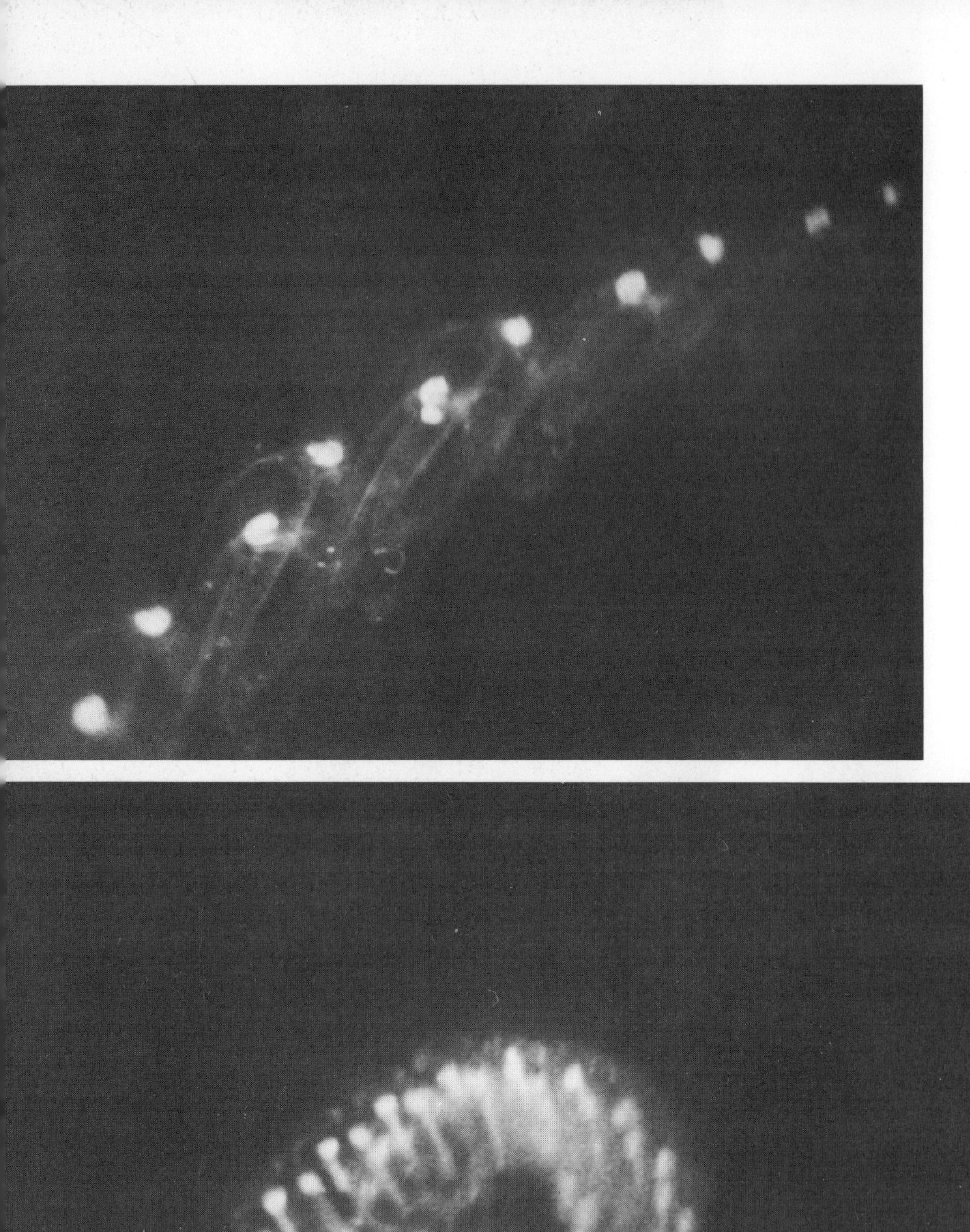

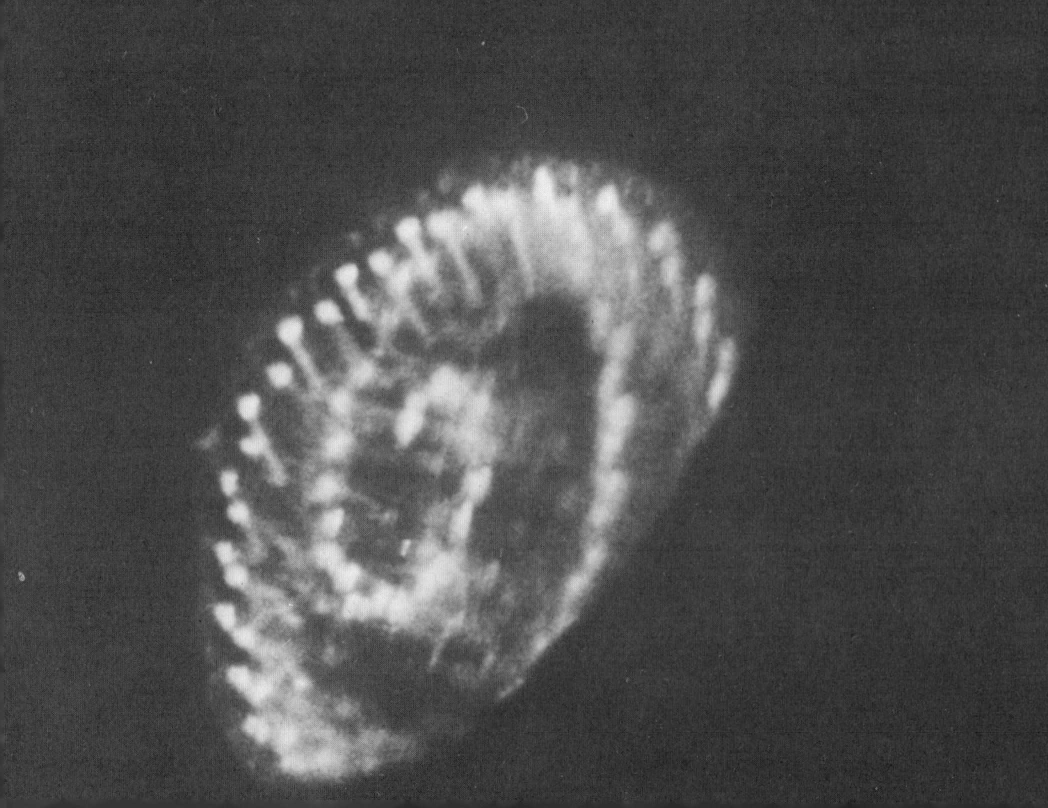

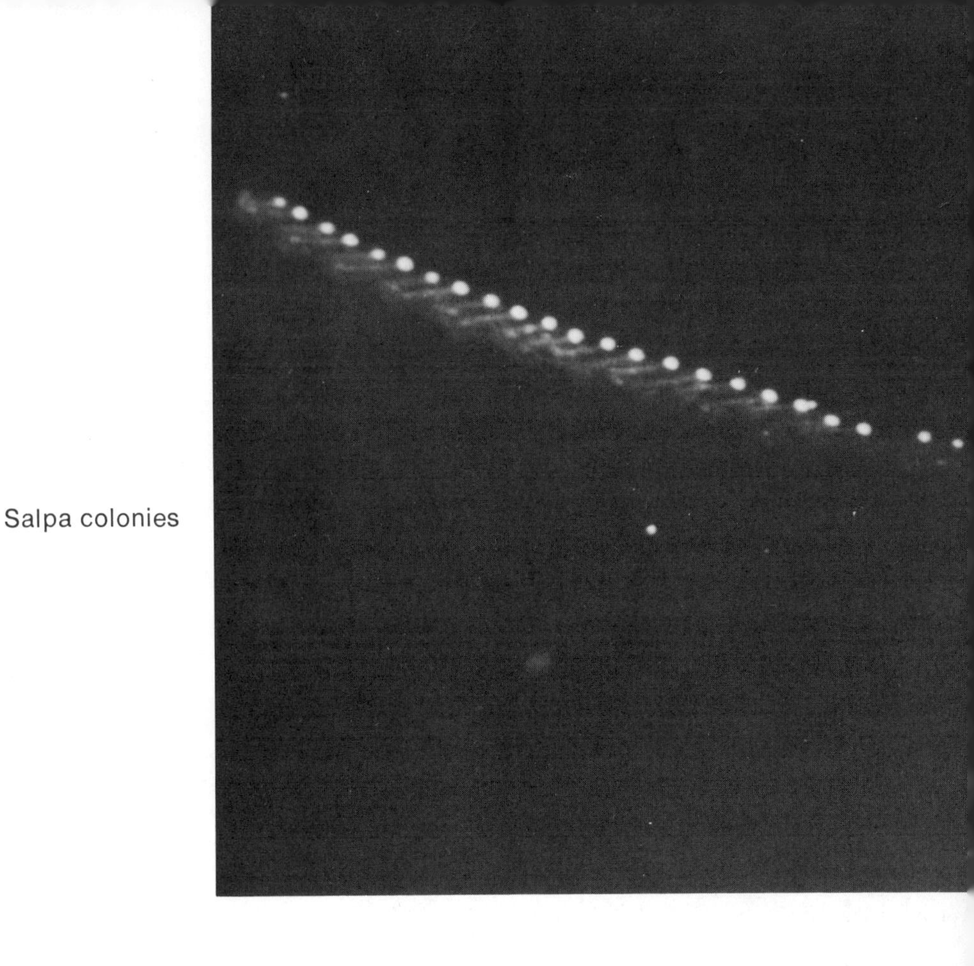

Salpa colonies

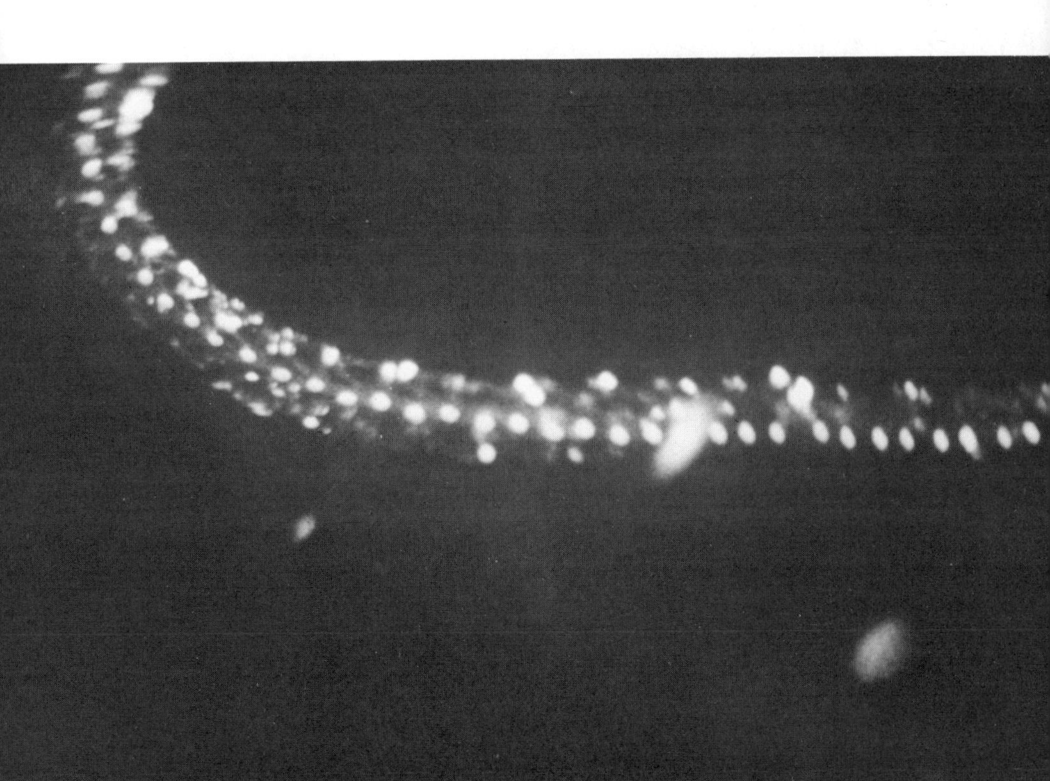

USCGC COOK INLET
PORTLAND MAINE
UNITED STATES COAST GUARD
1790
UNITED STATES COAST GUARD
1790

End of the mission:
(Left, top) The *Ben Franklin,* just after surfacing behind the *Privateer;* at right, the *Cook Inlet*
(Left, bottom) The members of the mission crew landing at Portland, Maine. Left to right: Piccard, Busby, Haigh, Kazimir, May, Aebersold
(Below) The *Ben Franklin* and its crew being welcomed in New York harbor

PART ONE

Mesoscaph *Auguste Piccard*

1

FIRST DIVE IN THE LAKE

The lake was absolutely calm except for the wide wake trailing behind our launch. The light haze in the atmosphere softened the horizon, and the Bay of St. Sulpice, toward which we were moving at an easy 10 to 12 knots, had somehow the look of the tropics and reminded me vaguely of a Pacific atoll. For the great experiment about to take place, the lake had put on a bit of a maritime air.

"There it is, a thousand meters straight ahead of us," the helmsman of the launch said to me.

"It" was the thing everyone was talking about—the mesoscaph. For more than a year it had been in the news; every week, every day, some paper discussed it; always with great respect, often with a large number of errors as well. There had been announcements of its conception, articles about its birth, development, and growth, then descriptions of its emergence from its parental home at the foot of the mountains. Like two rows of honor guards the Dents du Midi and the Dents de Morcle had saluted its first steps; a special train had welcomed it at the doorstep; to create it, mountains had had to be moved, and a bridge had to be shifted aside to make way for it. And then there had been the descriptions of its baptism, a great and solemn occasion; after that, lest it become tired, it had been towed to its new home below Lausanne. At each stage there has been cries of amazement, for from the first it had belied predictions, even the most pessimistic. People expected it to be small; it turned out to be big. They thought it would be gray or yellow; it was white. They said it would not be ready on time, and here it was, two months before the "opening."

At this point we heard a huge, muffled sigh, like that of a sperm whale after it has sounded. There was no one on the bridge. Around about were a few boats with microphones, with flashbulbs popping—and with functionaries too, recognizable from a distance by their bald spots shining in the late afternoon sunlight. The exhalation continued, while the mesoscaph yielded obediently to the weight of the water rushing into its ballast tanks, expelling the air from them and producing the rumbling sigh. Slanting gently forward, it was about to enter, for the first time, completely into its element. For an instant its conning tower of aluminum and Plexiglas hesitated; the eye of its television camera rapidly swept the horizon and came to rest, as if in embarrassment, on two clusters of barrels, which resembled the piles of gasoline cans that disfigure the landscape. These unsightly buoys had been haphazardly assembled at the demand of certain experts who had never been to sea and who unquestionably had never had their heads under water; terrified at the thought of resting on the floor of the lake, they had insisted that the mesoscaph be suspended from barrels to keep it a certain distance above the bottom.

After a few more seconds' hesitation the vessel began to sink; first the entire bridge disappeared, then the conning tower with its two antennas, and finally the flag with its white cross of martyrdom, hope and peace—this flag that would descend to a depth of 5 meters in the muddy waters of the Flon, but which four years before had happily gone down almost 11,000 meters beneath the rolling, crystalline surface of the Pacific. The flag disappeared; the barrels remained.

For a quarter of an hour those barrels were the cynosure of all eyes. Their importance was immense; on them depended all the hopes of the Exposition. They supported the weight of the mesoscaph, of course, but also the vastly greater weight of the thousands of worries and problems, past and future, the thousands of tiny intrigues already being hatched; they supported all those in whom the Committee of Organization had suddenly placed its trust—the engineers, experts, superexperts, various amateurs and certain wholly serious men as well. Also, in a chiefly wine-growing district, these barrels had a symbolic value; to all who were in any way responsible for this first dive, every barrel was the equivalent of an expert. If it split, the mesoscaph would go to the bottom, all expectations wrecked, prestige wiped out. But what constituted an alliance between the barrels and the men enclosed in the hull 10 meters below, the thing that lent them a certain kinship, was not their content or their capacity but the fact that they were empty. Their supporting power, their efficiency, their worth—these they owed to emptiness.

But a barrel is not supposed to split; inspected, calked, varnished, it must be ready at harvest time to receive the fruit of the earth. And these barrels like all others were true to form: to the last minute they bore their share of the weight. And when the mesoscaph emerged, everyone had the feeling that the real heroes of the day were those barrels.

You may ask how it came about that I was in a launch, a mere witness of this dive instead of an active participant. Well, that is another story, of which I shall relate only certain episodes, a story that began in 1953 on the return from another dive, at the mouth of the Gulf of Naples.

2

THE ORIGIN OF THE MESOSCAPH

It was a beautiful autumn evening. My father, Dr. Auguste Piccard, and I were sitting on the rear deck of the tug *Tenace,* which was running at low speed with the bathyscaph *Trieste* cautiously in tow. Two days before, the *Trieste* had reached the bottom of the Tyrrhenian Sea, a depth of 3,150 meters. The dive had proceeded "normally"; it was the sixth dive for the bathyscaph, and we had had the good luck to come to rest on a soft, sandy bottom and to make observations under conditions that, while not the best possible, were quite respectable. It was the first time that human eyes had been able to inspect the sea bottom at such a depth. But what work, what complications, to organize that dive in the bathyscaph! It practically amounted to a mobilization. We had brought aboard almost 100,000 liters of gasoline lent by Esso Italiana. Four tank cars and their four towing vehicles had been needed to transfer this precious cargo to our float in the course of one night. When everything was ready, a tug from the Italian navy had come to get us at Castellammare, the little town rebuilt on the ruins of ancient Stabiae which had been destroyed in A.D. 79 by the eruption of Vesuvius. At sea we were joined by the corvette *Fenice,* which would help us at the time of the dive. Altogether there were more than a hundred men—officers, sailors, engineers and technicians from the shipyards of Navamechanica where the bathyscaph had been assembled—a hundred men who were essential to the success of the operation but who at the time of the actual dive would remain, at least so it was hoped, on the surface to await our ascent.

Now the little squadron, consisting of the corvette, the tug, the bathyscaph, and a motor launch, was returning to the Gulf of Naples; everyone was satisfied, for the mission had been successful. But we knew very well, my father and I, that an operation of this sort, costing so much money, could not be repeated as often as was desirable for scientific research. Isn't it a natural part of the role of the scientist to find in every experiment, whether successful or not, the seeds of new experiments, new ideas, and new possibilities?

The sea was magnificently calm. We were running at about 6 knots and only a slight vibration told us that in a few hours we would reach shore. In the distance Mount Faïto, and farther off Vesuvius, loomed up through the evening mist. The sailors were preparing for their return to the base, putting things in order, closing lockers, writing letters which would be on the first train to Naples, Sicily, or the Piedmont. The slowly setting sun was just approaching the horizon. I remember the moment as though it were yesterday.

"With modern kinds of steel or even with Plexiglas," my father said, "one hull could be made lighter than water that would still let us descend to respectable depths—intermediate depths. If it's lighter than water, we don't need a float. A vessel of moderate size would do, and a hoist capable of bearing a few tons' weight could take us to the locality of the plunge and put us in the water. Each dive would be an easy operation and relatively inexpensive; oceanography would have much to gain from such a small submarine."

As he talked, more ideas came to my father's mind, and by degrees the vessel took shape. "To make it go down, since it would be lighter than water, we could, for example, equip it with a propeller in the vertical axis, somewhat like a helicopter. Also this would give us the advantage of total security; if the motor or the propeller broke down, the vessel would automatically come to the surface, since it would always be lighter than water. Of course, a submarine like this could not descend to very great depths, but that's not the point. How many totally unexplored cubic kilometers are there in the ocean at a depth of less than 1,000 meters? It is not necessary to make deep dives all the time—3,000, 11,000 meters, and such. Some of the most useful, important, and fruitful work from a scientific point of view can be done at tens or hundreds of meters beneath the surface. To be sure, this little submarine could not be called a bathyscaph since that name by its very etymology (*bathys* means deep and *skaphē* means a light boat) is reserved for vessels that go to great depths. What I would like," my father went on, "is to call this new vehicle a 'mesoscaph,' signifying that it is a vessel for intermediate depths."

It was on that day at that sunset hour at the mouth of the Gulf of Naples, between Capri and Ponza, that the mesoscaph was invented. It was October 2, 1953.

3

PRELIMINARIES

The project developed rapidly. In my laboratory at Castellammare I made the first sketches and experimental models. It soon became clear that an elegant solution to our problem would be to use two spheres instead of one, as the *Trieste* was designed. This would give us enormous advantages: greater stability and buoyancy, the space to carry a considerable array of measuring instruments, and, most important of all, a large independent range, thus automatically reducing operating costs. At this stage we envisioned a mesoscaph designed so that under normal conditions it could remain in the water and proceed under its own power to a dive site, without needing a ship to carry it or a tug to tow it. We were convinced that such a vessel could be built and operated economically, relatively speaking, that it would unquestionably be useful for oceanography, and that we would not have too much trouble getting it financed.

It was on the last score that we ran into difficulties. Though scientific circles were obviously eager to use this new means of investigation, they were for the most part quite unable to come up with sufficient financial assistance. I knocked at many doors which I had entered earlier while we were building the bathyscaph. Everywhere I met with the same response: "The mesoscaph? Oh yes indeed, very interesting. Why don't you ask Society X for help? Right now they are looking for something to back. Failing that, apply to Foundation Y." The greatest generosity always consists in recommending you to somebody else. We had to find someone who understood what "scientific investment" meant and also had adequate funds at the ready. In general, it can be said that this concept is foreign to the average European. In the United States it is familiar, even solidly established, but it has already had time to become embedded in regulations, controlled by huge foundations and omnipotent centers of research, and these have had to ring themselves round with protective networks that often make them completely inaccessible to individual researchers.

After carefully planning the construction of the mesoscaph and drawing up as precise and modest a budget as possible, I presented myself one day at the offices of one of the great American foundations. I was cordially received and seated in a comfortable armchair behind a large desk. After a brief exchange ("I know your country: how do you like America?" and so forth), we got down to the serious question, the only one of importance in the eyes of that organization.

"How much is your budget?"

"Half a million dollars," I replied.

"I beg your pardon? Didn't you say that your vessel was to descend to 1,000 meters, would have a speed of 8 knots, and would carry a number of observers aboard?"

"Yes," I said, "but—"

"You can't construct such a vessel for so ridiculous a sum. You'll need at least five million dollars. Don't ask for half a million. Nobody will take you seriously, and you'll get nothing at all."

Nevertheless the mesoscaph as planned could be constructed for that amount, at least in Switzerland. But this is one of the dramas of American research. Small endowments for modest ventures interest no one, while large sums for very large projects are reserved for months or years in advance.

It turned out that even the United States Navy was no longer interested in the mesoscaph because it had its eye on other projects (now in process of completion) which, it was then hoped, would perform similarly to our submarine. Also, the famous "Buy American" act was constantly cited to me. According to this, the United States government could not buy abroad matériel that existed at home or that could be built there. When the United States Navy had bought the *Trieste,* the bathyscaph with which we later made the dive to almost 11,000 meters and which later still was to play an active role in searching for the wreck of the *Thresher* (the nuclear submarine lost in 1963) and of the *Scorpion* (lost in the summer of 1968), the contract had expressly stated that the government had made sure that production of bathyscaphs in the United States was insufficient and that therefore an exception to the "Buy American" act was justified. At the time there were exactly two bathyscaphs in the entire world. One was the French navy's FNRS-3 which had been built in collaboration with my father, Dr. Auguste Piccard, and had made use of the sphere from the FNRS-2 built in Belgium; and the other was the *Trieste*.

Meanwhile, never forgetting our mesoscaph project, I was continuing to conduct dives with the *Trieste*, in Italy up to 1957 and in America from 1958 on. From San Diego was organized the series of dives in the Marianas Trench that attained successively depths of 5,000 meters, 7,000 meters, and 10,916 meters. These dives were the prelude to a second oceanographic campaign in the same region with new descents to great depths.

On my return to Switzerland I immediately went to work on the future mesoscaph. As I no longer had my laboratory at Castellammare, my first move was to create a new one in Lausanne. Fortunately I found a vacant apartment, not very large but perfectly suited to the project. I installed first of all a high-pressure laboratory: the pressure chamber itself was made by modifying—to only a slight degree, by the way—a 16-inch shell that had been given to me by the U.S. Navy. I must say that when this shell arrived at customs it caused a certain amount of stupefaction in an office accustomed to handling cigarettes and automobiles but not 400-millimeter shells. For a minute the officer in charge was plunged in gloom: a 400-millimeter shell? An American shell? Was this not contrary to Swiss neutrality? Would the Constitution have to be amended before deciding the matter? Would it be necessary to convene Parliament, to go to the people, to demand new elections, perhaps?

But with admirable good sense the officer in charge asked me what use I planned to make of this shell. When I made it clear that the shell was not to be exploded either in Switzerland or anywhere else but was to serve merely as a piece of laboratory equipment, the political-administrative breach was instantly healed. For laboratory equipment there was a schedule of duties; and as everyone knows, when there is a schedule the door is open for importation. What would the duty amount to? Oh nothing much, one franc per kilogram. Yes, but my shell weighed almost a ton. We then proposed to bring it into Switzerland as scrap iron.

"I agree," said the obliging customs officer, "but the regulations specify that scrap iron must 'show signs of rust.' "

Feverishly we opened the packing case, and everyone including the customs officer went to work searching for a trace of rust. The shell had come from America; it had been admirably brushed, greased, cleaned, polished, wrapped; it gleamed like a mirror. I would never have dared take the initiative in discovering any trace of rust on it. Beneath its solemn and austere manner Swiss customs possesses immense reserves of common sense; one of the officers, more perspicacious than the others, equipped, if I remember correctly, with a powerful jeweler's loupe, made a discovery.

"Here, look," he cried suddenly. "A spot of red, a spot of rust." By saying this, he did more for scientific research in his country than many committees meeting in plenary session.

"Then what is the tariff?" I asked.

"Nine hundredths of a franc per 100 kilos."

And so for 90 centimes my shell weighing one ton passed the frontier, without percussion cap but with head held high.

Modified at the Ateliers de Construction Mécanique de Vevey, provided with an inlet for thirty electric cables and with a Plexiglas porthole similar to those in the bathyscaphs, the shell became a splendid pressure chamber. Fed by a high-powered electric pump, it could simulate depths far surpassing those at the bottom of the great oceanic trenches. A whole series of interesting experiments was carried out. But I still had not found anyone able to finance the building of the mesoscaph.

4

THE EXPOSITION

The event that would decide the issue, however, had been in preparation for quite a while. There is a theory that Switzerland holds a great national exposition on the eve of each world war. If the country should be destroyed, archaeologists of the future would easily be able to reconstruct a model of Swiss society—everything would be in one place, arranged, analyzed, explained, summarized, and beneath the ashes and the ruins one would find skeletons of cows, blueprints of watches, the formula for milk chocolate, and plans for the future of the teaching profession. In point of fact, Switzerland has not engaged in a general war for a century and a half and has never been reduced to ruins. But "how can you tell?" as the people like to say.

With the international situation critical, our

economy lush, our industry in high gear, a new exposition had been decided on for 1964. Why 1964? Because that would be twenty-five years since the last one and fifty years since the one before that. This would facilitate the statisticians' calculations.

Many people, as a matter of fact, wondered whether it really was a good idea to hold an exposition at that time. The economy did not require it: the country was rolling in money, advance-order books were crammed for years ahead; everyone, even the laziest, had his chance at a share in the national wealth; the population had increased by 15 percent through the addition of foreign labor because Swiss workers had become too fussy, too delicate, and there were too many kinds of jobs they were unwilling to undertake. Everywhere trains ran, planes flew, and automobiles collided on the highways; bankers were refurbishing their façades, often adding a veneer of marble. Was it really worthwhile to spend millions to stimulate an economy overflowing with resources and riches? To increase the flow of tourists, which already exceeded the capacities of the country anyway?

Furthermore, even the idea of an exposition seemed old-fashioned, obsolete. With the great expositions such as that in Paris in 1889, or Berlin or London, which had caused so much talk and excitement, the principal purpose had been to concentrate in a single city examples selected from the entire world; thus a mountaineer had a chance to see a taxi, a hotel, the inside of a factory, a famous painting, an amusement park; or a city-dweller could see a stable, a cheese dairy, or a model pasture; a white man could see the model of an African village, or a Chinese an American Indian's wigwam. All this, meanwhile, surrounded by flags whipping in the wind vigorously enough to stimulate all the passions of nationalism that raged in the nineteenth century—love of country and hatred of one's neighbor.

By the middle of the twentieth century things had changed radically. Millions of tourists roamed around everywhere in Europe, blocking the roads, causing insoluble headaches for hotel keepers; in the course of their travels these potential visitors to the Exposition were seeing 95 percent of what it was proposed to show them: electrical transformers, high-voltage lines, amusement parks, museums, giant locomotives, superbly muscled bulls, and model kitchens. That's not the point, replied the champions of the Exposition. We will show them other things too; we will let them see the soul of the country, the intrinsic virtues of its citizens, the moral force of its population. We will create the "Swiss way," which will reveal the evolution of thought, the white cross of the nation's flag, and the future of the Confederation. We will have stunning motion pictures, automatic cannon that will leave our future soldiers to their daydreams, a high tower to "see the view," and unequaled scenery. Needless to say, there will be an attraction—a special attraction worthy of the twentieth century, worthy of Switzerland, worthy of the world that will send us visitors by the millions. Switzerland will become the cynosure of the entire universe; its future, its neutrality will be assured forever.

An attraction?

Yes, but what attraction? On this problem the wisest gentlemen had discoursed at length around more than one round table. Someone had proposed—O miracle of the imagination!—a tower 300 meters high.

Just consider, a tower with its feet at the border of the lake and its platform higher than the church of Saint-François!

Someone objected because so tall a tower might interfere with airplane traffic. The rebuttal to that was that the Eiffel Tower did not do so (and for good reason: airplanes are forbidden to fly over Paris.)

Nobody considered another possibility of constructing a tower which would be 500 meters tall but placed in a pit 500 meters deep. There would have been numerous advantages.

The tower would not interfere with airplane flights nor with the view of nearby places; being completely protected from the wind, it could be of light construction and would, moreover, be cheap; its top, by definition the most interesting point, could be observed on an equal footing by the public, eye to eye. The tower could also be seen from above, something altogether unique for a tower. Besides this, the Exposition could boast of having created not only the tallest tower in the world but also the biggest hole to put it in.

There were many other proposals, but not one of them seemed able to win the vote of the majority.

It was about this time that I had the pleasure (as it was then) of entertaining one of the directors of the future Exposition at dinner in my home. This good fellow was even a sort of relation of mine, one of those distant relations by marriage that involve no consanguinity and can quite easily be denied if that seems indicated. After dinner we talked about the Exposition.

"You have proposed something to do with the lake?" my guest inquired, pulling desperately at his pipe.

"Yes," I replied. "Why don't you commission me to

build the mesoscaph? The vessel could be used later for scientific research, and during the Exposition it would permit a large number of visitors to make a tour of the bottom of the lake."

"Wouldn't it be dreadfully dangerous?"

I gave him a long account of submarine navigation. I explained the principle of the bathyscaph and the mesoscaph as compared to that of an ordinary submarine. I pointed out to him that beneath the surface there was no traffic problem.

"I've been thinking of this project for a long time," I said. "I've already carried out in my laboratory a whole series of experiments bearing on its construction. Naturally there's no question of using the mesoscaph in its first design for the Exposition. In fact, the vessel would be very different; in place of one or two spherical cabins that could accommodate only a few people, I would have a cylindrical cabin similar to that of an airplane; at either end of the cylinder there would be a hemisphere. On both sides of the hull I would have a row of portholes, behind each porthole a comfortable chair."

"Then you're planning to put windows in it?" he asked in dismay.

"Of course. You couldn't get thousands of visitors to go down without lettting them see the bottom of the lake."

"Thousands of visitors? But what dimensions do you plan for the mesoscaph?"

"I'd consider that forty persons is a minimum. If the vessel makes fourteen dives per day and continues to operate for the duration of the Exposition, that will make a maximum of 100,800 persons. You could hardly expect to attain this figure, for there would be idle days, bad weather, days reserved for maintenance; but some tens of thousands of persons could certainly take part in these dives."

My cousin the director was thunderstruck. This was point one gained. He would convince the steering committee of the Exposition and within a few weeks everything would be settled. That was October 1961, two and a half years before the opening day, which was more than enough time to build the mesoscaph and make the necessary tests.

However, it took fourteen months for these good people to reach a decision, and God alone knows how many meetings and how many committee reports. When on December 10, 1962, the Exposition finally approved the construction of the mesoscaph the vote was, it appeared, unanimous.

Objectively considered, the careful weighing of the pros and cons of my proposal is understandable. In a strict sense of the word my project did not fit the definition of an "attraction" for such large crowds as were expected. At that time they were counting on selling some ten million tickets (later, when the budget of the Exposition began to swell, they counted on selling eleven million, then twelve million, and finally there was talk of sixteen million visitors, which made it possible to have a continuously balanced budget on paper, at least until the day when the gates closed). In the best of cases less than 1 percent of the visitors could find a place aboard; what about the remaining 99 percent? Admittedly a considerable percentage would quite mistakenly think themselves too old to take part in a dive, and of sixteen million admissions, several million would be from Lausanne and the surrounding country who would not dive more than once. There might also be a lot of people with claustrophobia who would not dare take a dive. Even so, it was obvious that many people would have to be turned away at the ticket counter of the mesoscaph and that it would not be universal in appeal as had been hoped. In any event, however, hundreds of thousands of visitors would see the mesoscaph maneuvering in port and diving and reappearing near Lausanne, would hear the whip of its flag and the song of its siren. And they would see the red symbol of the Exposition on its white turret.

Despite all the palavering, despite the local interests with their narrow-minded arguments showing a total incomprehension of the problem (one member of the committee asked me if there was not the danger that the mesoscaph, when rising to the surface, would leap some tens of meters high out of the water like the fountain at Geneva, killing everyone during the very first dive), despite all such nonsense, it is certain that the final decision came largely from a profoundly noble feeling. I have always remained grateful to the Exposition for *that* decision.

"One" would have the opportunity of showing what "one" could do: the Exposition would be a showcase exposed to the whole world (at least so it was thought—actually many people avoided Lausanne in the summer of 1964 for fear that the Exposition would add to traffic problems). The reputation of Switzerland would be enhanced; national industry could develop in a new direction. *The mountaineer will have a chance to dive in the ocean.* What a headline for the sensational press!

Everyone had an interest in the Exposition's proving to be large, beautiful, successful, worthy of a people proud of their past and confident of their future. The federal authorities supported it, so did the canton of Vaud; the city government rubbed its hands in anticipation of the large profits that would easily assure winning the next election. As for those in charge of the Exposition, each one would have to find a new position once the gates of Vidy were closed, and this would be quite a problem. In a period of economic prosperity, no great industrialist, no important businessman is much inclined to leave his own field and throw himself into an adventure that is risky, and temporary as well. So there was the danger that management would consist mainly of second-rate people who did not have much to lose anyway and who might hope to establish new careers with the Exposition as a personal springboard for their future. This is fairly close to what happened, but the springboard was not, so to speak, solid enough and showed serious cracks, thus considerably diminishing the distance of the final leap. The Exposition was directed by a triumvirate of two apprentice dictators and one architect; beneath them at the base of the pyramid a large number of honest folk who came to work at the Exposition inspired by pure idealism, believing that they ought to become the mirror of their country's grandeur. I will have occasion to talk later on of this mirror. When the Exposition closed, the triumvirate found acceptable re-employment without too much trouble. One became a grocer, another an ironmonger, and the architect stayed in architecture.

But let us not anticipate. The date is December 10, 1962; the directorship of the Exposition has decided in favor of the mesoscaph and has charged me officially with its construction. There is barely a year to accomplish this.

5

DESIGNING THE FIRST MESOSCAPH

As a matter of fact, I had not waited for this day to undertake preliminary studies and have the first blueprints made by my office force. With the agreement of the Exposition, which assumed the financial risk and thereby anticipated the final decision, I had already placed certain orders, prepared a number of contracts, cleared the ground, engaged personnel. Collaboration with the Exposition proved excellent at that time: everyone did his part, and I was allowed the greatest liberty both in organizing the work and in the application of technical ideas.

From the day when I had decided that the cabin should be cylindrical and not spherical, the general appearance of the vessel had been determined: outside, it would look like a submarine; inside, like a commercial airliner.

For what depth should it be constructed? The basic law of all floating vessels requires that their weight should equal that of the water displaced; the greater the volume of displacement the more the vessel can and must weigh. In the case of the mesoscaph the volume was determined in general by the arrangement of the interior as required by the space necessary for passengers. To avoid the risk of claustrophobia, I had decided not to compartmentalize the interior but to have a cabin as large and open as possible; in short, to give the passengers at the very moment they entered a feeling of spaciousness. For the cylindrical hull, I had chosen an exterior diameter of 3.15 meters; from the point of view of transporting the vessel, it would be advantageous not to exceed this size. After consultation with the technicians of Swissair, we estimated that we should leave a minimum of 70 centimeters space between chairs. The total length was determined by the number of passengers, twenty on each side, a figure that had been thought reasonable. All this determined the exterior dimensions and consequently, as had been seen, the weight of the mesoscaph.

The weight of the entire vessel had to be divided in general terms among the following elements: the hull, ballast tanks, keel, batteries, propulsion system, emergency ballast, various accessories, and the passengers.

The weight of the ballast tanks was determined by their volume, the weight of the keel by the stability desired for the vessel, that of the batteries and the engine by the performance and range desired. The maximum weight of the passengers was also known, as was that of the emergency ballast; the accessories could be estimated. The sum of these weights had to be deducted from the total weight, the difference giving the weight of the hull and, for a specific steel and method of construction, the depth to which it could be taken. In round numbers the mesoscaph was to weigh 165 tons: the ballast tanks (that is, the exterior reservoirs which, filled with air, assured easy floatability on the surface and, filled with water, allowed the submarine to dive), 5 tons; the keel, 17 tons; the batteries, 20 tons; the motor and its accessories, 1 ton; the emergency ballast, 5 tons; the passengers and crew, 3 tons; there remained 80 tons for the hull. Taking into account the stiffeners—those rings designed for an especially important role in the struggle against buckling, one of the classic forms of destruction of a hull under exterior pressure—this weight of 80 tons made it possible to have the principal steel wall of the mesoscaph 38 millimeters thick. Calculations showed that such a hull could withstand pressure to a depth of about 1,500 meters. This does not mean that our mesoscaph could descend to that depth, but that if it did descend, for example, to 300 meters, the depth of Lake Geneva, it would have a coefficient of safety of 1,500 over 300, in other words, of 5—assurance that it could descend farther than a nuclear submarine with equal safety. The principle of the coefficient of safety is worth a little further analysis, for it is often misunderstood.

First of all, if the critical depth is known (that is, the depth at which the weight of the water will crush the submarine), why need one remain so far this side of it? Why not descend, let us say in our case, to 1,450 meters? A precise definition of the safety coefficient may give the answer.

This coefficient is the ratio between the pressure which according to calculations or experiments would probably result in crushing a vessel and the pressure to which it is subjected.

The word "probably" ought not to astonish the reader: despite all the precision of calculation, all the knowledge gained from experiments by the engineer, there is in a structure as complex as a submarine a whole series of data which are not precisely known, especially at the moment when manufacture begins. There may be irregularities in the thickness of the plates, or even in their composition, in the composition of the metal used in welding, in the electrodes. There are always geometrical irregularities, differences between the specifications and the finished vessel; the hull is not perfectly circular and the degree of imperfection is never uniform for the *whole* hull. Likewise, in a dive there are unknown factors: the density of the water, especially due to variations in temperature or salinity which can change from one place to another, and most significant of all, that density can cause a submarine to descend suddenly to a greater depth than was foreseen. For example, for a large nuclear submarine running at 25 knots, a difference in salinity of 1 per thousand, common in the ocean, can add 8 to 10 tons of weight and cause a descent that may carry the vessel beyond its safe depth in less than a minute. It is clear that as more and more calculations are made and more and more experience is acquired, the coefficient of safety can be progressively reduced. It is also clear that only the breakup of a submarine could make it possible to establish *exactly,* though too late, its "coefficient of safety," which is the same as saying that the coefficient of safety is only a formula for the insecurity of the whole apparatus.

In classical mechanics this coefficient of safety, whenever human lives were at stake, was fixed in the neighborhood of at least 4; actually methods of structural analysis have made such great progress that the general tendency is to reduce the coefficient. In a relatively simple structure—the spherical cabin of a bathyscaph, for example—a security coefficient of 2 would be ample. In the mesoscaph, the first of a new structural type of submarine, designed for tourists, it was prudent to have a high coefficient of safety; calculation and experiments on models and precise measurement of the deformation during dives showed that in Lake Geneva this coefficient would always remain between 4 and 5 and that therefore the risk of accident was practically nil.

The function of the ballast tanks has already been mentioned; in a conventional submarine, these tanks often surround the whole rigid hull. Here, we had to avoid having them interfere with the view; therefore they were arranged laterally above the portholes on both sides of the hull. Another fundamental advantage of the arrangement was that it gave the mesoscaph a constant, positive stability.

A conventional submarine (modern nuclear submarines are no longer in this class) has two sorts of stability: on the surface, stability of shape; under water, obviously, stability of weight. Omitting details at the moment, it should be noted that when the dive begins, the conventional submarine loses its stability of shape before attaining that of weight; as a result, for a few moments the submarine is theoretically unstable.

In order not to capsize at this point, it moves forward, supporting itself on its ailerons as an airplane does on its wings. A purely static dive without horizontal motion could lead to capsizing. However, the arrangement of the ballast tanks in the mesoscaph gives it a permanent, positive stability; it can perfectly well dive and re-emerge without horizontal motion, a maneuver essential in a submarine designed for tourists. Also, in case of engine trouble, the mesoscaph can rise to the surface without risk of capsizing.

The ballast tanks, arranged in six sections on each side, had a total volumetric capacity close to 24 cubic meters. It is this capacity that allows the vessel to float fairly high when it is on the surface, but this considerable volume has to be filled with air each time it emerges. Forty-two cylinders of compressed air, each containing 50 liters under a pressure of 250 kilograms per square centimeter, supplying therefore more than 500,000 liters of air, are sufficient in principle for twenty dives and reasonably sufficient for the nine dives a day that were planned, even taking into account the air used for the variable ballast tanks and for the siren.

The principal function of the keel is to assure stability of the mesoscaph while under water (and to contribute to its stability on the surface). This keel is a relatively light assembly of steel plates welded to the hull and forming watertight compartments in which ingots of lead can be placed, 14 tons in all. Thanks to the lead, the distance separating the center of gravity from the center of buoyancy was of the order of 21.5 centimeters. In simple language, this means that the mesoscaph would slant less than 2 degrees when a man of average weight moved from one end of the hull to the other.

The batteries presented a series of problems of which only the principal ones will be touched on here.

Although it is possible—I shall return to this apropos the mesoscaph *Ben Franklin*—to place electric batteries outside the hull, for various reasons I preferred in this case to have them inside the mesoscaph. Of course this made a series of precautions necessary for the safety of passengers and crew, in particular to avoid diffusion of gas, especially hydrogen, which could mean a disaster if it reached a critical percentage in the interior atmosphere of the mesoscaph. So the batteries were in the bottom of the hold in watertight tanks, and a system of ventilation and precise controls with alarms and automatic safety arrangements was installed aboard. At the time these batteries were the most powerful in any non-military submarine and also, let it be said in passing, the most powerful bank of batteries ever constructed by their supplier, the Electrona Company of Boudry (Neuchâtel). Six hundred kilowatt hours were supplied in the course of a day of nine dives, to the engine, to the sixty-one external floodlights and to the electrical equipment aboard. The batteries were sufficient to give the mesoscaph a range of 200 kilometers at 4 knots, or 90 kilometers at 6 knots.

In building the first bathyscaph, the FNRS-2, Dr. Auguste Piccard had the idea of putting the driving motors of the submarine outside, subject directly to the pressure of the water. A light tank and a bath of oil in which the motors ran prevented short circuits and grounding. This system saved weight, gained space in the interior of the bathyscaph, and eliminated the stuffing boxes which can be very troublesome at high pressures. I had planned to follow the same system in the mesoscaph, but in the end I placed the driving motor inside the hull as is done in conventional submarines. It turned out that the interior spaces of the mesoscaph were big enough, its lifting force sufficient, and the passage of the driving shaft through the hull not particularly difficult to arrange for depths that were small in comparison with those where the bathyscaph reigned supreme. We installed a type of stuffing box in use on German submarines, and we improved on it considerably so as to give the mesoscaph all the safety required.

The motor itself was a direct-current engine of 75 horsepower especially built for us by Brown Boveri and Company; its maximum speed was 1,500 revolutions per minute, with a resulting propeller speed of 300 revolutions per minute.

In addition to these security features described and many others, a particularly noteworthy measure was the 5 tons of emergency ballast in the form of iron shot held in place by magnetic valves, following the system also invented by Dr. Auguste Piccard; this would let the submarine rise to the surface at the instant of the pilot's command, and also in case of a breakdown in the gates' electrical circuit.

The figure of 5 tons was arrived at in the following manner: The mesoscaph has "trim tanks" by which the pilot can on the one hand give the submarine any desired longitudinal inclination and on the other can, before beginning a dive, increase or decrease its weight as required by the weight of the passengers. One can imagine, however, that the commander, after a dive with a class of schoolchildren weighing a total of 1,500 kilograms, might next be diving with a group of the "100 Kilogram Club" weighing over 5 tons. This would no doubt prompt the commander to be especially prudent, but one can conceive that he might forget to lighten the mesoscaph and to empty, or empty sufficiently, the trim tanks. Then the mesoscaph would be tons too heavy. Depending on the circumstances, the pilot would be forced to get rid of several tons of ballast to regain the surface. We put the figure at 5 tons to give a greater margin of safety.

Now for a summary of the principal fittings:

> The bridge, which protects some of the pipe connections and some of the electrical cables as well as the instruments on the outside and on top of the hull. Also it provides a base of operations for the crew when the mesoscaph is on the surface.
>
> The tail, which gives the mesoscaph an efficient, hydrodynamic shape and places the propeller far enough astern to assure efficient operation.
>
> The conning tower, which allows entrance and exit from the mesoscaph during bad weather without risk of rain or waves getting into the interior of the hull.
>
> The diving planes and the rudder, hydraulically activated and electrically controlled from the pilot's post. A system of servo motors, developed for us by Allgemeine Elektricitaets Gesellschaft (AEG), allows the pilot to maneuver the planes with ease and to know at all times what position they are in.

> The two variable ballast tanks, small reservoirs (250 liters each) on the outside and resistant to pressure, which make it possible to increase or decrease the weight of the mesoscaph to an exact degree, no matter what depth it may be at.

A submarine, as already pointed out, should normally weigh at most the same amount as the water it displaces. In practice, certain exceptions can be made, as when a naval submarine is attacked by an airplane or fears such an attack and must dive as quickly as possible. To do this it increases its weight by taking on a great quantity of water, a maneuver that makes it descend very rapidly. When it arrives at the desired depth, it must get rid of the excess water or run the risk of continuing down to a dangerous depth. In any case, increasing its weight in this fashion means considerable risk, and it is up to the captain to decide whether the game is worth the candle. As a matter of fact, this technique has been so perfected by submariners that it is a part of the training system and is used on maneuvers, in general exercises as well as in time of war. Nevertheless it sometimes happens, alas, that a submarine fails to rise.

To avert this danger, I had arranged for our submarine always to dive a little lighter than the water and to be taken down by the joint action of the propeller and the diving planes. Thus even a breakdown of the motor would have no serious consequences, since the mesoscaph, light as it was, would automatically rise again to the surface. It follows that the equilibrium had to be meticulously adjusted before departure; theoretically the craft should never be heavier than water except when resting on the bottom, specifically in the course of dives for scientific purposes.

In actuality the operators of the mesoscaph, all experienced professional submariners, did not adopt this technique; they generally dived heavier than water, and had a perfect record during the whole time in Lake Geneva.

In any case, the exact weight of the vessel had to be known before departure; but since one could not expect all the passengers to conform to the average weight tables, it was decided to weigh them individually before each dive.

From a layman's point of view, the mesoscaph had many interesting characteristics.

Entering through a wide, welcoming door to go down a steep ladder, the tourist (in this case the average Swiss, for he constituted the majority of the visitors) was surprised by the attractive appearance, the warm colors, the open space inside the mesoscaph. How different from the interior of the hull of the ordinary submarine of the maritime powers, which the privileged visitor may enter on a national holiday, or the anniversary of some battle such as Jutland—a day when defense secrets can be exposed to the eyes of the public since the spies are all on vacation. In an ordinary submarine everything is different, with narrow doors and low, obstructed corridors, pipes barring the way or busy sailors whom one has difficulty recognizing as the future accomplished heroes of the coming war. There is also a sharp odor of burned oil, of acids or of fried potatoes, which catches in one's throat and remains until one has quit the ship. In the mesoscaph, on the contrary, there is nothing of that sort. The wide, long cabin is well lighted, has many portholes and two rows of twenty armchairs apiece, upholstered in reddish orange and provided with seatbelts as in an airplane.

Above the seats are banks of handsome blue cylinders containing compressed air for the ballast tanks and the siren; the television screens are here too. Beautiful lighting—absolutely brilliant. From one end of the submarine to the other are reinforcing steel arches which give the interior a robust and reassuring aspect: the arches resemble those of a Roman church and the hull itself makes one think of the ribs inside some marine monster created by Walt Disney. At each end of the mesoscaph is a technical compartment. Forward is the pilot's post, to the layman much like that in a big jet plane: three seats facing forward, at center front a series of dials, push buttons, screens, voltmeters, signal lights and control levers, all arranged opposite the seats in such a fashion that a single pilot could, if need be, operate the vessel by himself. But the three seats indicate that the vessel is handled by three men: a commander, a pilot, and a navigator or engineer.

Behind the passengers, protected by grills and panels, are the driving belts, their flywheel, and the driving shaft; to port, the motor itself, the electrical center with its numerous control dials and fuses; to starboard, the hydroelectric station permitting operation from a distance of the access doors, the diving planes and the rudder, as well as the flood valves controlling the dive itself.

6

THE ACTUAL CONSTRUCTION

Construction began with enthusiasm. As soon as the agreement with the Exposition was concluded, I selected the principal suppliers. The hull was to be manufactured by the firm of Giovanola Frères of Monthey, a mechanical workshop specializing in the production of hydroelectric conduits and consequently able to manufacture the rigid hull of the mesoscaph without too much trouble. The bridge, the ballast tanks, a part of the tail, and the battery cases were to be made by the Ateliers de Construction Mécanique in Vevey. This company had already done work for us, notably the completion of the interior of the bathyscaph sphere which descended to 11,000 meters in 1960. In Vevey there was a fine engineer, Edouard Pugliese, whom I had known for a long time; he took charge of the general coordination of the enterprises between Vevey and Monthey. Excellent relations were immediately established between us and the personnel of the two companies; everyone set to work courageously, knowing that this was his contribution to the success of the Exposition.

Naturally enough, there was occasionally some skepticism. Time had run by since the first sketches, and when the first plate for the hull was rolled the date for the launching was less than a year off. How could we complete construction in a year? How could we make the trial runs in the lake in two months? How were we to bring this prototype to perfection in so short a period? Just the same, everyone really felt confident. A submarine, however great its novelty, is never anything but an assemblage of machines that can be put to other uses, if not all together, at least each in its separate role. Then there was the example of the *Trieste* which was often cited though without taking into account the fact that there is a world of difference between a bathyscaph built for a few dives at great depth and a mesoscaph destined to make thousands of dives in a lake on a rigid timetable.

Nevertheless the schedule was met. Giovanola performed its first miracle by building the hull in six months; the second miracle—in reality more important—was the precision, the perfection, of the finished work; practically every part, every detail, was a masterpiece of workmanship and came within the tolerances demanded by our calculations and experiments. From then on, my principal collaborators, particularly Erwin Aebersold and Christian Blanc, two young technicians of the first order, and I myself spent a good deal of our time in Monthey, overseeing the operations and directing the work. When the hull was completed, the general responsibility of Giovanola came to an end, but the mesoscaph remained in the principal erecting hall so that the final fitting out could be done there. The accessory equipment arrived from all over Switzerland and some of it even from abroad, and little by little the vessel took on its final shape.

This period was one that I shall never forget. When I had taken part in the building of the *Trieste,* my father and I were perfectly free to organize our work in any way we liked, and no one could presume to give us any directions whatever. On the other hand, our budget was worse than limited; we continually had to refrain from purchasing some piece of equipment that proved too expensive; we always had to economize until it hurt, as by making a mere letter take the place of a trip of inspection. The final result was none the worse for it: the *Trieste* performed perfectly its numerous missions, but the amount of work accomplished suffered considerably from these financial restrictions.

With the Exposition enterprise, there were no such problems. We had a general budget from the beginning, and credits had been established. True, time did not allow us to go into every detail, and the theoretical budget was in fact exceeded, particularly because the managers of the Exposition, as the work progressed, insisted reasonably enough on a vessel not only of fine quality but one that incorporated practically all the technical and esthetic advantages possible. In practice, I ordered the equipment, received the bills, signed them, and sent them to the Exposition for payment. No builder could dream of a more ideal situation for accomplishing his work. In reality, this was not due simply to generosity and understanding on the Exposition's part but was largely the result of a very heavy-handed administration and of a kind of inertia—once the operation was set moving in one direction, it could never be swerved from its course. Like the hippopotamus which crushes everything in its way when it is rushing toward the water, the Exposition plunged forward without much reflection, having only one thing in view, the opening date.

At the beginning of 1964 the mesoscaph was practically finished. Its conning tower of aluminum and Plexiglas, also built at Vevey under the direction of William Nicole, a remarkable engineer and specialist in light metals; its delicate and elegant tail, its powerful propeller; its diving planes which gave it something of the look of a shark; especially its rows of portholes—all this gave it the appearance of being alive. Everyone had his imaginary picture of it in action—breaking the surface of Lake Geneva, disappearing into its depths, secure from the violent wind squalls that can transform the mirrorlike lake into a very bad-tempered little sea.

Just before Christmas there had been a moment of near panic. The hull had been constructed in three sections, each section heat-treated to eliminate the final welding stress, but the three sections, once welded together, would not fit into the oven which was nowhere near large enough to accommodate the whole mesoscaph. This contretemps came about suddenly because solidity tests carried out on plates of the same quality, and subjected to the identical treatment as our hull, revealed that the two final bands of weld, which had not been reheated like the rest of the hull, did not have the expected strength. Many technical discussions followed, and appeals to a number of welding and metallurgical experts; we had to decide what type of reinforcement could be added to give the hull uniform strength despite the two bands of weld. From the start Giovanola proposed that small plates be welded onto the hull to prevent the steel from buckling in the two critical zones. This solution had nearly all the technical advantages required but the disadvantage was that it would hold up the work, so it was thought, for at least a week or two. Local thermal treatment of the welds was also considered, either directly with the torch or through a high-frequency device.

I myself could not agree to this solution, for I feared that local reheating might cause new strains in the adjoining areas. Finally, it was proposed that a large furnace be built rapidly, to reheat the whole mesoscaph, but this was a fantastic notion since "rapidly" would have meant at least three or four weeks, not counting the delay caused by all the work that would have to be redone. Most of the fittings were already in place, the painting was finished, and this could not have withstood the thermal treatment. If time had permitted I would have had flanges made to bolt the various sections together instead of welding them, but this solution, which I was to use for the second mesoscaph, would in the present case also have taken too much time.

While the theoreticians were busy discussing the problem, undeniably crucial for the future of the mesoscaph, the workshop under the direction of the foreman in charge, Charles Weilguny, had secretly chosen its own solution and was preparing the reinforcement material in order to be ready to weld it in place as soon as the decision had been officially made. Weilguny had already done us invaluable services and

God knows how many more he was to do later on. When we asked him to proceed with the operation as quickly as possible, he astounded everyone by saying, "It will be done tomorrow." Those fifty strips of sheet metal, 450 kilograms in all, are certainly among the finest Christmas presents I have ever received. As for Weilguny, who lived in his workshop, who gave himself body and soul to his task, who drove his men into accomplishing their share of the miracles he wrought—who knows whether it was not also one of his finest Christmases? Alas, such miracles seldom repeat themselves, and no one can defy the laws of nature with impunity; Weilguny, who put together our first Swiss submarine and four years later assembled the second mesoscaph, the future *Ben Franklin* of the Gulf Stream, literally killed himself at his work. The day we began our drifting dive in July 1969, he was the victim of a stroke from which he did not recover: I only learned the news a month later when we returned to the surface; it threw a pall over our expedition.

Conditions of work were not always easy. Monthey is, as European distances go, a long way from Lausanne. The road was bad in winter, crowded in summer, and the constant shuttling back and forth was exhausting. At the beginning of autumn, it is true, our base was transferred to Monthey in a semipermanent way. Everyone found lodgings there, in most cases in private homes. But the atmosphere of Giovanola was especially favorably for our undertaking. The extraordinary degree of cooperation between management and personnel facilitated our relations with the company itself; top priority had been given to the construction of the mesoscaph, and the red tape was actually reduced to the barest minimum. A hundred times, a thousand times, it was necessary to open a storeroom already officially closed; every weekend some overseer was called upon in his free time to come and open the factory if we thought it necessary. And how many workmen gave up their vacations or postponed them simply to hasten the work!

For our part, we had formed a skeleton crew that had the last word in matters of construction and worked closely and effectively with the people at Giovanola. Despite differences in nationality and background, in ages and in character traits, these groups soon became one group, with excellent morale and a true esprit de corps.

Nevertheless, as I have said, work conditions were not easy. The overall assembly of the mesoscaph was carried out during the winter of 1963-64. The workshop at that time was not heated; the noise in the erecting hall was often intolerable; the hours of work—day team, night team—were longer than an average man, even one well trained, could ordinarily stand up to over a long period. Yet the arduous routine seemed not to have any ill effect on the health; cases of grippe did not slow the rhythm, so great was everyone's desire to finish on time.

I had set February 27, 1964, for the launching of the mesoscaph, which gave us just two months for trial runs and training of the crew before the Exposition opened on April 30. A week before that date, the submarine was to leave Monthey and be transported to Bouveret where a special launching ramp had been constructed.

This trip of about 20 kilometers had been prepared for with extreme care. (Wasn't this, I wondered, the first time a submarine would make a journey by train before plunging into its element?) It had not taken long to decide that we should use the railroad, and all the expertise of the CFF (the Swiss railroad) was called upon to carry out the transfer operation successfully. As soon as the decision to build the mesoscaph had been made, I had got in touch with the CFF and was assured that the trip was theoretically possible. When I stressed the fragility of some of the instruments, the great size of the vessel—in short, the importance of what was at stake—the official in charge of difficult shipments replied nonchalantly: "Oh, you know, it doesn't matter a bit to us whether we're handling a box of macaroni or a submarine."

The load, as a matter of fact, even when placed on very low trucks (which chanced to have been made some years before by Giovanola) exceeded the maximum permissible height; by just about 5 centimeters it was too tall to get under one of the bridges, fortunately only one. There was no alternative but to cut the bridge and lift it 6 centimeters to let our submarine get by. The government of the Canton, the traffic police, and the transportation authorities all agreed; the bridge was cut and lifted and the mesoscaph passed through.

The trip was certainly spectacular. It took place at night so as to cause a minimum of inconvenience. The day of departure from the factory had the look and feeling of a public holiday. Work practically stopped, since all those who had taken part—and who had not at least fetched a stick of weld or a rivet, or applied a stroke of the paintbrush?—wanted to be there at the dramatic moment when the mesoscaph left; for the personnel it was rather like a baptism.

When everything had been done and made ready (loading, removal of all equipment that protruded beyond the diameter of the hull and the ballast tanks, final retouching of the paint work, enlargement of the door of the factory), we awaited the locomotive which was to come and fetch the mesoscaph. There was a moment of anxiety when a rumor suddenly circulated that the submarine could not leave because the railway guard had not received his "way bill." Finally that was cleared up, the two sides of the great door opened, and slowly, very slowly, the convoy got under way. As it passed through the door of the workshop the submarine sounded its siren for a whole minute, its bass voice bidding a last adieu to those without whom it could never have existed. The shipping department handed me a certificate of clearance for the departure of "one mesoscaph (1)."

7

DRAMA IN LAUSANNE

The journey from Monthey to Bouveret took place without incident. At 15 kilometers an hour on the straight stretches and 10 kilometers on the turns, the convoy made its way through the night, spotlighted by the flashbulbs of photographers and the floodlights of television men. Many workers and assistants followed in cars and on bicycles. At one o'clock in the morning the mesoscaph arrived at its destination. With a final grinding of brakes it was placed in position for the launching.

Some hours later, at the crack of dawn, the whole crew reassembled around the mesoscaph, ready to put back everything that had been removed for the trip—diving planes, conning tower, and other accessories. Six days remained to get ready for the launching. The weather during this interval was favorable, but cold; in the morning we had to wait for sunup before climbing onto the bridge, for the fine layer of ice that had formed in the night made work difficult and dangerous.

This was the point when the drama of the mesoscaph—for there was one—started to unfold. Not too definitely disturbing yet; nevertheless there was menace in the air.

Shortly before, one of the directors of the Exposition had come to see the mesoscaph in the workshop at Monthey. The visit turned out to mean a terrible blow for him, a revelation. Up until then he had not really believed in this project to which he had himself voluntarily subscribed. When he found himself face to face with the mesoscaph, his responsibility suddenly filled him with dread. No technician, he imagined a whole series of possible dangers, any one of which, if it became real, could result, so he believed, in sending him straight to prison. Suddenly all the words that he had seen in orders, invoices, or technical reports—hull penetrators for cables, driving shafts, Plexiglas portholes, strength of steel, yield point, critical depths, regenerated air, and so forth—took on in his mind a nightmare aspect. It seems that from that day on he could not sleep, which did not improve his disposition. At the next meeting of the steering committee of the Exposition, he unburdened himself to his colleagues, wishing to diminish his responsibility by sharing it with others.

Naturally enough, the others wanted nothing to do with the so-called problems, and since someone had to take upon himself this unexpected and crushing responsibility, they decided to appoint immediately a "committee of experts." When in any gathering someone proposes the creation of such a group—a scapegoat committee, so to speak—nobody dares say no, for in case of an accident the whole responsibility would fall on him. No one had the common sense to let the builders complete their work, to allow the progressive trials to proceed, and to collaborate as from the beginning with that first-rate engineer and specialist from Trieste, Benvenuto Loser, builder of submarines for over forty years, engineer-in-chief at Monfalcone Yard, whom I had known well since 1952 when he helped my father and me in the building of the *Trieste.* No, here were my administrator-politicians creating a committee of "neutral" experts, obstructing the work and imposing on us control by consultant engineers who had never seen a submarine. When they finally added a real submariner to the group, it was already too late. Here came our experts on tiptoe, suddenly appearing in Bouveret like sorcerers out of the underbrush, less than a week before the launching, waving a letter that gave them full powers, proposing theories that would make even laymen shudder, and in the name of Swiss prudence challenging everyone's good faith, putting all things in question and declaring that the mesoscaph could not dive.

In the midst of frightful confusion, each one wanted to disapprove of something specific in order to sound his own negative note and, above all else, to temporize. As Victor Hugo would have said: "You could tell by the bald heads scattered around the table that experts were sleeping here!" One fine day my director approached me with the air of a man struck by lightning.

"You have forgotten something."

"Ah?"

"Yes, the torque."

"What do you mean, the torque?" I asked in stupefaction.

"Yes, yes," he repeated. "If the mesoscaph proceeds under water with one plane jammed in the 'up' position and the other jammed in the 'down' position, this will produce a torque." (Without knowing the meaning of the word he was obviously fascinated by it.) "The mesoscaph will then spin as it advances and everyone will be killed!"

The poor fellow had believed in the lucubrations of the "neutral" experts. This group had not taken the time to calculate that the hypothetical accident it had thought up would not only fail to make the mesoscaph spin in the water but would cause it to tip by less than 4 degrees! But the idea had spun its way into my unhappy director's head and remained forever fixed there.

If I mention these examples, it is to emphasize how disastrous the arbitrary imposition of controls can be when these are suddenly applied by total incompetents (whatever their qualifications in their own professional fields), and when the controls are more administrative and political than technical. Here it was as disastrous as it could possibly be. It discouraged the personnel who had been keyed up by months of slaving away and thought they were arriving at last at the goal of their labors; it paralyzed the work with claptrap talk completely contrary to common sense. The echoes of the conflict were bound to reach the general public and very quickly stimulated a veritable war in the press. The Exposition had first aroused these passions by issuing a pessimistic statement which alarmed public opinion. The committee of experts, given absolute authority, threatened to resign if it were not completely satisfied; it demanded a number of modifications in the mesoscaph which, if carried out, would have postponed the beginning of its operations and its service to the public until the Exposition had closed. There was even an attempt to prevent the launching, though this was finally authorized, but with the proviso that I must shoulder all responsibility for any failure that might ensue, a failure they were doing their best to make inevitable. Of course I was more than willing to bear my responsibilities, and it was precisely this that disturbed my adversaries.

In fact, the whole affair was not so much a technical argument as a human issue of the most banal kind: one of my collaborators, the thirteenth, call him Judas, a competent-enough engineer but a wicked man, longed to take my place. He made the cause of the experts his own, took to groveling before them, disowned his own work to the point of self-accusation, brayed his newly learned lessons of subservience before the committee, promised everything that was asked provided only he be allowed to parade the gangway of the mesoscaph for the duration of the Exposition.

But before the storm became audible to the public, the launching of the mesoscaph was finally authorized for the day we technicians had chosen. It was the last untroubled day for the mesoscaph, and the ship certainly looked handsome—all white with a bright orange line at the level of the bridge, which emphasized its great length, and a flag waving above the conning tower. Completely assembled and ready to take to the lake, the mesoscaph was about to be baptized and given to the water. The ceremony was half magnificent, half rustic; that is to say, it was a complete success. The fire that smoldered beneath the ashes was quiescent; the directors of the Exposition were there in formal dress; a Protestant pastor and a Catholic chaplain bestowed blessings on our submarine. The president of the Exposition said a few words and so, for that matter, did I; the sponsor then proceeded with the baptism. The sponsor was Mme. Auguste Piccard who gave the mesoscaph the name associated with the very origins of the exploration of the deep sea—Auguste Piccard.

There had been some reluctance about using the traditional bottle of champagne, but finally it had been agreed upon since it was a part of seafaring custom. The moment the bottle broke on the bow, the powerful siren of the mesoscaph sounded, the cable that still held it to the land was loosened, and the *Auguste Piccard* gently slid into the water. Flourish of trumpets, applause, flags, banners, songs, explosion of flashbulbs, presentation of oars by the local lifesaving society—that ceremony lacked nothing in fanfare. It had been exactly a year since construction had begun in the workshops of Giovanola.

On that same day the mesoscaph was towed to Lausanne. Its driving system could not yet be operated because the bearings for the driving shaft were made of lignum vitae and had to be soaked for several days in water before they could be permanently installed. We had favorable conditions for the crossing from Bouveret to Lausanne, arriving at nightfall opposite the location of the Exposition, marked so far only by workshops. We were counting on getting to work again next morning and proceeding with the first trials.

That was when everything fell apart. It is impossible for me to give the details of all the intrigues going on at that time. Their first result was the dismissal of the entire team of the mesoscaph except Judas, and the formation of a new team that, unfamiliar with the submarine, had to begin again from the beginning and lose an incredible amount of time, thus leading to the second result: a delay of two and a half months in the opening of the mesoscaph's doors to visitors at the Exposition. My own collaboration with the Exposition became impossible; this was what the committee of experts were working for when they made technical demands so outrageous that they knew I would never be able to accept them, and which they themselves were willing to give up after my departure. Fortunately, to the original Swiss group was added a French builder of submarines, who should have been made president of the committee of experts. But his role, useful though it was, unfortunately was effaced by the dictatorship of the Exposition.

For weeks the unhappy experts beat about the bush, not daring to make a move, losing precious time by demanding futile and absurd tests. Their recommended modifications were of three kinds: some silly and dangerous, some completely useless, and others so minor that they could be classified merely as final improvements. Later on the experts gloried in these last, which would have been made in any case, no matter who was in charge. One example of the first category was the elimination of the safety microswitches that I had had placed in the door circuits to prevent the mesoscaph from diving if the doors were not completely closed and locked. This was a new safety measure, which does not exist even in the conventional naval submarines. There the only security—the best in the eyes of the military—is obedience and discipline on the part of the personnel. "Aye aye, Captain," accompanied by a sharp clicking of the heels is preferable to any kind of safety circuit. The result has been that, out of 160 naval submarines sunk accidentally—that is, not as an act of war—between 1851 and 1960, 26 (16.3 percent) have been sunk by water rushing in through a gate or door that stayed open when it should have been closed. But this was something my brave experts did not know and, during the whole period of the Exposition, the dives began by order of the commander after he had been told by a subordinate that the doors were shut. That no accident occurred showed how well disciplined and trained the personnel was. But after the Exposition—when the influence of the experts had doubtless diminished—the safety switches were replaced! This case was the most flagrant among many, but my point is not to stir up old quarrels, but merely to provide the reader with the reason that I watched the first dive of the mesoscaph from a launch. It would be a pity to postpone the explanation for posthumous publication.

8

JOURNALISTS FOR BALLAST

A day finally came when the Exposition thought it could make a great impression on the public mind. It announced clamorously that the press had been invited to go on the first dive of the mesoscaph. Since I was practically out of touch with the Exposition, they wrote asking me to be present at these initial dives and to talk about them to the successive groups of journalists who would take part. I replied that the mesoscaph was not ready to dive, that no one in the new crew knew the vessel well enough to be able to guarantee the success of its first dive, and that it was premature, to say the least, to invite the press. Therefore, I had to refuse to participate actively or even to be on board. The real point of this was, of course, that they were afraid of a failure and wanted to be able to hold me accountable.

It turned out to be a beautiful day. About a hundred journalists had applied. Most of them had been present at various stages in the construction of the mesoscaph and, knowing the care with which it had been put together, had no hesitation now about going aboard. Sound of trumpets, flourish of bugles, flags—the atmosphere of an operetta dominated the ceremony. At ten o'clock twenty journalists were taken aboard for the first dive. The doors were shut, the flood valves opened. My Judas strutted in a style to match his new contract.

"Ladies and gentlemen, you are about to descend, the flood valves are open, we are descending, we are descending. Ah, how magnificent!"

"What's the depth now?" a reporter asked.

Judas looked at the manometer; somewhat embarrassed, he admitted that they were not yet very deep but nevertheless it was fine.

"What's the depth now?" another reporter asked.

"To tell the truth, we're still on the surface, but we're going to descend, we're going to descend. You'll see, it will be magnificent!"

After a quarter of an hour it had to be admitted that the mesoscaph, though it floated, would not sink.

"Hello, hello, mesoscaph here! Tender, tender! Mesocaph here!" cried a slightly agonized voice, coming from the radio on the tender, from another at the command post in the harbor, as well as from the listening post that I had installed for this occasion on my balcony 10 kilometers away. It certainly was worth the trouble. I recorded the conversation.

"Attention, tender! We are too light. Send three more journalists!"

"Message received. Three journalists are being sent."

The ballasts were blown. The mesoscaph, not having left the surface, took on a new attitude somewhat higher in the water. The doors were opened, three journalists entered, the doors were closed, the flood valves were opened.

The small adventure occurred all over again for the three newcomers.

"Ladies and gentlemen, you are about to descend. The flood valves are open, we are descending, we are descending. Ah, how magnificent!"

"How deep are we?" asked one of the newcomers.

"To tell the truth, we're still on the surface, but we're going to descend."

And so the story repeated itself. After several unsuccessful attempts, they asked not for three more journalists but for journalists by the kilo, without further concern for the press, which they felt was already losing interest.

"A hundred and fifty kilograms of journalists!"

"Three hundred kilograms of journalists!"

"On their way!"

But each time the mesoscaph refused to go down. After two and a half hours, the cabin was jammed with reporters. Despite all efforts, no more could be crammed in. Yet the mesoscaph did not descend.

This time it was a tearful voice that spoke from the radio:

"Listen, this won't work. We can't go on. I don't know what's wrong. We shall have to give up."

Once more the ballast tanks were emptied, the door was opened.

"All ashore!"

At this command most of the passengers rushed toward one of the exits. The tender happened to go rapidly past the bow of the mesoscaph at that moment, making a wave break on the bridge already slightly tipped by the passengers massed on one side of the vessel. It suddenly struck the newsmen about to disembark.

Nerves, strained by two and a half hours of unsuccessful and grotesque trials, almost broke at that point. A cry of "We're sinking!" caused near panic. But the passengers' instinctive recoil righted the vessel and all was well. Everyone went ashore. The dive was put off until afternoon.

From noon until 2 p.m. the experts inspected the mesoscaph, trying to find out what had kept it from descending.

Finally it occurred to them to glance at the electric indicator which showed the level of the ballast, something they had forgotten to do before attempting to dive. And they saw—that there was *no ballast aboard.* Too light by 5 tons, the mesoscaph had a right to stay on the surface! Later, the ballast was found at the bottom of the harbor, and a sailor assured me that an "expert," whose name he gave me, had dumped the emergency ballast without realizing what he was doing. Five tons of supplementary journalists would have been required, something not easily found even in a country where the press is free and flourishing.

In the afternoon the ballast was replaced and the dive took place to more or less general satisfaction.

This experience showed that a sailor, even a fresh-water sailor, cannot be improvised from just any expert or apprentice engineer. It was high time to form a crew, or rather three different crews who could relieve one another and assure a quasi-permanent service. The answer seemed to lie in hiring true submariners, but there too things did not go without a hitch. The Exposition addressed itself first of all to the French navy, which having got wind of the intrigues, refused to collaborate in any way. An appeal was made to retired officers. Many applied, which was looked upon with disfavor in Paris, and it was clear from the first that those who volunteered to help operate a fresh-water submarine would incur the wrath of the admiralty. Despite this, there were a number of applications, some of them semiclandestine; the amateurs did not dare enroll openly and one of them went so far as to present himself publicly as a former German officer!

9

THIRTY-THREE THOUSAND PASSENGERS UNDER WATER

Everything was finally arranged. A French crew supplemented by an Italian crew worked admirably, in really first-rate fashion. The mesoscaph, I believe, fascinated them. What a difference between a naval submarine, a blind brute which could do nothing but sow terror and death (and for which nevertheless submariners could have such a passion, for one can love a monster as well as an angel), and the mesoscaph, delicate, elegant, gracious, supple, as easy to control as an airplane in fair weather and built for the sole purpose of taking innocent tourists for a ride.

Very quickly these foreign crews mastered the problems which, for them, posed hardly any difficulties at all. Nine times a day, six days a week (Monday was reserved for inspection and maintenance), more than seven hundred times during the Exposition the mesoscaph left the harbor of Vidy and dived to a depth of about 100 meters to rest upon the bottom of the lake. More than 20,000 passengers made the dive and, thanks to the installation of closed circuit television, could follow and admire the navigation on the surface and the execution of their own dive on the television screens. Underwater they were fascinated by the precision landing on the bottom, always in the same area, where everyone could see the marks the mesoscaph's keel had left on its preceding dives. These marks, long furrows in the lacustrine sediment, sometimes made one think that the tombs of the experts had already been dug in the mud.

On the surface the mesoscaph was accompanied by a small tender. This was a motorboat able to carry three or four persons and was also used to help maneuver the mesoscaph inside the harbor. Once a French officer, commander of the mesoscaph and by rank a captain, next grade to admiral and qualified to command entire fleets and cruisers of the *Richelieu* class, wanted to take the helm of the tender. One of the Exposition functionaries stopped him because he did not hold a motorboat license. It was so preposterous that the captain thought at first it was simply a bad joke. It was no joke, but rather evidence of a virus which had infected everyone involved in the management of the Exposition. The French officer had to threaten to resign before he was finally allowed to operate the tender.

Strife had now reached a point where the law had to be called in to decide who was right and who was wrong. A tribunal of arbitration was formed according to an excellent formula of Swiss law, by which the Tribunal's final decision is absolutely irrevocable.

I shall pass over the hearings at which the stature of the experts was revealed as so minuscule as to be hardly visible; the expertise and the counterexpertise; the memoranda and the countermemoranda. The tribunal, held in secrecy like all such tribunals of arbitration, was presided over by a true judge, in this case a judge of the Supreme Court, the highest court in the Swiss Confederation. When the day of final judgment arrived, the meeting was held in the Federal Tribunal in Lausanne. In a somber room with splendid paneling, seated in elevated chairs, were the technical arbitrators and the legal arbitrators along with the assessors who surrounded the judge. Everything was very handsome. The door opened and the representatives of the Exposition entered, thoroughly confident that they would win and that justice was for sale to the highest bidder as it had been when the Borgias set about buying their tiara. But Justice was no longer for sale, and the decision hit them like a club. They went out with heads down, tails between their legs, and were seen no more.

Once the fair was over, final accounts had to be rendered. The result, as expected, was less than brilliant. The furnishings were liquidated, the cake was divided, the crumbs were sold. The proceeds were not enough to balance the books. The situation was so bleak that there was reluctance about revealing the accounts to the sovereign people. The mesoscaph for a while saved the situation: it was announced that the submarine was still a part of the assets of the Exposition and that the accounts could not be rendered until it was sold. There was, however, no reason to think that it could be sold quickly. After some months of inactivity it was put back into service for occasional tourists. It made some four hundred additional dives, which brought to 1,100 its trips to the bottom of the lake, with a total of more than 33,000 passengers—figures that attest to complete success. Thanks to the veteran French and Italian sailors, the operation took place without a hitch, without an accident, without an incident of any kind. To the best of my knowledge, only two trips were interrupted as a cautionary measure, each time for minor causes which would not in any event have entailed serious consequences.

10

MESOSCAPH FOR SALE

When the Exposition was over, the directors started negotiations for the sale of the mesoscaph. Unfortunately people who might have had the ability to sell a herd of cows had considerable difficulty in deciding a sales program for a mesoscaph that would satisfy everyone. At the outset they counted on getting over $2 million—that is to say, more than the cost of construction, the launching, the home port at Vidy, and the whole operation, including the salaries of personnel and all incidental expenses. So they loudly announced that they already had a hundred offers to buy and that all they had to do was to choose the best and worthiest bidder. Apparently they could not agree on the choice; since their first cry of anticipated victory had frightened off many serious amateurs who believed the sale had already been concluded and did not even put in an appearance, they had to lower the price.

"Come one, come all! Less than two million dollars for the first Swiss submarine, a unique opportunity, an exceptional bargain! Eight million francs! No one? What am I bid?"

By 1965 they were down to $1,500,000. The policy of mythical or declining bids worked wonders. By gradually decreasing the price they could attract the attention of amateurs who presented themselves in greater and greater numbers. Italians, Frenchmen, Egyptians, Israelis, and of course Americans were in the ranks. This time one could still seriously reduce the bids.

"No one bids a million and a half ? What am I bid? One million? Half a million?"

To judge by official announcements the buyers were jamming the doorways. The sale was thrilling. Obviously the price was still too high. By lowering it to $250,000 they would double the number of amateurs.

"And now, ladies and gentlemen, the first submarine of the Swiss navy! A quarter of a million dollars a trifle! One million francs! What am I bid? No bids?"

"Two hundred thousand dollars?" Together with certain friends I entered the ranks. The price was still too high.

"Two hundred thousand dollars? What am I bid?"

"One hundred thousand dollars?" A representative of a large American company deposited this sum on the table.

"One hundred thousand dollars—any other bids? None? Sold! One mesoscaph for one hundred thousand dollars."

Since the sum was far from what had been expected, various extraordinary rumors were put about, which led to the belief that the selling price had been ten times higher. Then it was announced that the price would never be revealed.

The final accounting was hurried over, the Swiss Republic permitting every liberty. Taxpayers have strong backs. Why rekindle a fire that has been put out? In any case, there will not be another Exposition before 1989.

Despite everything, the balance, technically and scientifically if not financially, was positive. A first mesoscaph had been built. This facilitated the building of the second, also constructed in Switzerland, the PX-15 or *Ben Franklin.* The first adventure showed clearly the human errors to be avoided. And in fact, as will be seen later on, an agreement with the American Bureau of Shipping, a neutral organization by definition, would have eliminated that dreadful question of responsibility, even if it could not have eliminated all the other problems.

O you who took the decision to build the mesoscaph *Auguste Piccard,* who assumed general responsibility for the enterprise and for its exploitation, you were great at moments, and I am not angry at you for your weaknesses and your pettiness, for after all your defects are only human. With the verdict of human justice already delivered on you, I believe that now a higher Tribunal might even acquit you.

PART TWO

Mesoscaph *Ben Franklin*

11

THE MEETING WITH GRUMMAN

One fine morning in 1965 I had a telephone call from Germany.

"Doctor Piccard? I am Marc Bailly-Cowell, representing Grumman Aircraft Engineering Corporation." (The firm is now called Grumman Aerospace Corporation.) "I am calling you on behalf of Walter Scott, director of Grumman's Ocean Systems."

I had of course heard the prestigious name of Grumman, the American maker of airplanes, both military and private, especially of a magnificent version of the executive plane called the Gulf Stream. I had seen one of these Gulf Streams in Florida, and I knew that Grumman had a base at Stuart, near Palm Beach. The company was also working on the lunar module for the Apollo missions.

The voice on the telephone continued:

"I wonder what you're doing at present. Would you like to work with us? Have you any free time?"

With my reduced staff I was busy on various laboratory researches, particularly developing plans and preliminary studies for three types of submarine. The idea that a great company might become interested in submarines instantly came to mind and I replied: "I can find time if you have something interesting to do."

"That's fine. I'll come to see you."

Two weeks went by without a word. Then one day Marc Bailly-Cowell arrived and we spent the day together. I showed him my laboratory, the researches under way at the time, and we agreed that our negotiations should continue in New York or, to be exact, in Bethpage, Long Island, near New York, Grumman's main office. I was invited there in February 1966.

The reason Grumman should concern itself with my work on submarine research is worth noting. Almost all the big American aircraft manufacturers—Douglas, North America, Lockheed, Martin, Grumman, and many others—had one by one opened up departments of submarine research, and for a good reason. The role of military aviation has been considerably reduced by the prospect of an atomic war. The course of World War Two was systematically determined by aerial supremacy, at first by Germany when the Allies had no effective aviation, next by England in 1940 when the Royal Air Force gained the upper hand, and finally by the Allies in 1944 and 1945 when their fighter planes and bombers were sole masters of the skies above the battlefields.

With the advent of the atomic bomb, the elements

of the problem changed radically. The victor is not necessarily the one who has the most bombs to deliver, and the weaker party may hope to destroy his adversary if he can get his bombs to the target first. The delivery vehicle is no longer the airplane, which is too easy to knock down, but the rocket, often carried aboard a nuclear submarine.

Thus American industry had to reconvert in part, for there was no longer any reason to produce or to foresee the need for massive aerial fleets of several tens of thousands of planes. The objection was voiced that local wars of moderate size might still occur, but military men have a way of taking the large view and they do not like little wars, except those that allow them to train their personnel and for these, industry, in America especially, can always deliver matériel enough and to spare.

The space programs partially took up the slack in the aviation industry, but oddly enough the public still did not believe in space as much as it believed in the sea. And in America the public is almost always master of the situation, since it is its opinion that is expressed through the press, radio, and television and it that keeps a firm hand on Wall Street, on the financial health of the country.

As a matter of fact, after World War Two the public became fascinated by the sea and oceanography because it needed something, and the sea was there. Space exploration was still a long way off, and perhaps, too, science fiction had overworked it to such an extent that few people really believed in it. Of course, space took its revenge, brilliantly.

In 1965 Grumman, wishing to diversify and to expand its activities into future prospects while keeping the majority of its personnel for aircraft production, was in the process of opening a submarine division, later to be called Grumman's Ocean Systems.

I presented to Grumman three different projects which I had been studying for some years, each of which had, naturally, its advantages and disadvantages. One was a small, light submarine, limited in range and inexpensive to build. That was rejected; it is of little interest today, for it has been superseded by other vessels. The second project a high-speed maneuverable vehicle, is still interesting, and I hope to be able to construct it sometime. The third was the one Grumman chose. This project, codified under the temporary name PX-15, was a mesoscaph intended primarily for the exploration of the Gulf Stream. Grumman, launching itself on the sea, wished in its first attempt to accomplish a master stroke.

The discussion that day was brief. It was followed, after a luncheon with the board of directors, by formal agreements drawn up by lawyers. These agreements envisioned a collaboration of at least five years between Grumman and myself, a series of researches together, and, most important the construction of the PX-15. In this respect Grumman immediately showed largeness of vision and practicality. In order that each of us might profit by the experience Giovanola had acquired during the building of the first mesoscaph, to keep our budget within reasonable limits, to profit by all the advantages of construction in Switzerland by a number of manufacturers I already knew, and also—why not admit it?—to re-establish a certain credit which the Swiss manufacturers of the mesoscaph had seen as slightly tarnished through the adventures of the late Exposition, I proposed that we also build the

PX-15 at Monthey in the Giovanola plant.

Grumman accepted these points of view, and it was thereupon decided to construct the new mesoscaph in Switzerland.

During the first months it was necessary to re-examine the problem and to establish the general lines of the project. I had been planning for several years either to use the first mesoscaph or to construct a new one in order to study the Gulf Stream. Here a parenthesis is needed: just why study the Gulf Stream?

12

THE GULF STREAM

There are many reasons for studying the Gulf Stream.

The sea is a bottomless pit of problems, questions, mysteries, doubts, unknowns, and challenges. Thousands of different investigations in 1.5 billion cubic kilometers of water might each hold the key that would lead, if not to an explanation, at least to a new problem, and this is in accordance with the philosophy of science today. However, some problems contain enough different factors to command more interest than most others. These are the problems that directly concern a large proportion of humanity, quite simply through their generality: one might mention the interaction of the surface of the water with the atmosphere (of fundamental importance for meteorology); the pollution of the sea; the function of the waves, the tides, the possibility of fishing at various depths, and many others. Among these "many others" is the Gulf Stream.

The Gulf Stream, of equal interest to America and to Europe, is an enormous field of study, relatively little explored. Enough of the preliminaries have been got out of the way so that each new detail, each new bit of knowledge, is directly useful because it can be integrated with what is already known. I shall not analyze in detail the origin of the Gulf Stream; this has already been done by others, especially by the scientists at Woods Hole, Miami, Fort Lauderdale, the oceanographers of the United States Navy, and many others who are specialists in this domain. But for a proper understanding of this narrative, some general information may prove useful.

The Gulf Stream is part of an important system of currents which constantly move through the North Atlantic Ocean and are directly connected with the currents of the South Atlantic and also of the Pacific. One can picture the Gulf Stream roughly as follows: As it leaves the Gulf of Mexico, it is rapid and hot, heavily charged with billions of calories which it has absorbed in the tropics. It swings around the Florida peninsula; confined between the American continent on the one hand and Cuba and the Bahamas on the other, it forms a true river in the sea, flowing northward at a speed of 4 to 5 knots. This speed, however, diminishes rapidly to 3 or 4 knots or less as soon as the Gulf Stream emerges north of the Bahamas and, taking advantage of the open ocean on its right, spreads more and more widely. Above latitude 30° N. and especially from Cape Hatteras at about 35°, it takes a more and more easterly direction and at the latitude of New York is plainly preparing to cross the Atlantic. There, much wider and consequently slower, it poses innumerable problems for oceanographers, the least of which is, perhaps, to say whether it still exists. There is no doubt that a transfer of water, energy, heat, takes place, but the scientists at the Woods Hole Oceanographic Institution who have studied the current for years, have found it so variable, so changeable, so capricious, that they have gone so far as to suggest that it has nothing to do with the original Gulf Stream but is a completely different configuration. According to classical terminology, however, which is still accepted at least in its general outlines, the Gulf Stream, when it arrives in the middle of the Atlantic, divides into two principal branches, one flowing north and the other continuing toward the east.

The northern branch divides to flow around Great Britain, one arm brushing the coast of France; there the two arms come together again in the North Sea and the branch enters the Sea of Norway. This it traverses, turning counterclockwise, then comes back along the coast of Greenland to the west of Iceland, and is finally lost in the cold waters of the Labrador Current, which disappears in the depths of the Atlantic along the northeast coast of the United States.

The eastbound branch makes straight toward France, rebounds from the French coast, travels south along Spain and Portugal, passes by Gibraltar where it sends a reconnoitering party into the Mediterranean, proceeds along Morocco, and suddenly, at the equator, turns westward. It recrosses the Atlantic, passes between Cuba and the Bahamas, and turns toward the north. At this point it is joined by another current that comes from the South Pacific. The latter current which has gone eastward, passing around the tip of South America at a distance, also rebounds from South Africa after following its western coast almost to the equator. At that point, it turns directly west, crosses the Atlantic, runs along the northern coast of Brazil, enters the Gulf of Mexico, and emerges south of Florida, where it joins the Gulf Stream as described. Then the cycle begins again.

According to official nomenclature, the part between the Gulf of Mexico and Cape Hatteras is called the Florida Current, and only the section extending from Cape Hatteras to the latitude of New York has a right to the name Gulf Stream; the rest, north of the equator, has the general and noncommittal name of North Atlantic Current. But in popular terminology the Gulf Stream is still the current that crosses the North Atlantic and brings to western Europe the heat that the sun gave it in the equatorial seas and that permits palm trees to grow in the south of England and wheat to flourish in the north of Norway.

It is perfectly obvious that such a mass of water, more voluminous than all the rivers of the world combined, in perpetual motion, growing warm in the tropics, losing its warmth in the boreal regions—in other words, carrying there calories from the south—has a tremendous effect on the meteorology of the western hemisphere, and consequently on customs, habits, character—in short, on western civilization as a whole. Modern science is demonstrating more and more the interdependence of all social and natural phenomena; there is hardly any limit to action and reaction, and the legend of the earthworm that fell in love with a star contains a whole philosophy: by its contemplative immobility the earthworm has a different effect upon the star than if it were in rapid motion.

Because of the Gulf Stream's importance to everyone—oceanographers, meteorologists, navigators, scientists in many fields, the military, and thereby politicians—I knew that an important new study of the current along the eastern coast of United States would be highly acceptable on that side of the Atlantic. In the choice of a scientific study—at least if one wishes to be completely independent from a financial point of view, which is extremely rare in our times—one must take into account the assistance that must be asked for: financing the equipment, loan of surface vessels, procurement of launching facilities, all the incidentals. For any one of these loans, from the moment a governmental department is approached in any democratic system, public opinion plays a major role: the press quickly takes hold of the project, criticizes and studies it, offers comment constructive or otherwise, and often gives the green or the red light to the government agency that will have to finance it. Thus the agency in question, whether an academy of science or a naval station, will always reach a favorable decision more readily if it thinks the project may be a popular one; in other words, between projects of equal scientific value you will more easily find financial support for the one that captures the public imagination. Many scientific projects of great importance have been eliminated because they came "before their time" or because too few people understood their true importance. Still, what could be more absorbing, more pertinent, than a new study of the Gulf Stream? Even as a youngster I had, like thousands of other students, been fascinated by this river in the sea that brings from the Caribbean islands, the cannibal country of the epoch of great discoveries, enough heat so that Switzerland is not entirely covered with glaciers. At least that was the belief at that time. More recently the theory has been partially revised; Professor Columbus Iselin of Woods Hole even believes that without the Gulf Stream the climate of Europe would be even warmer. So there is a great deal more work ahead before everyone is in agreement.

Historically speaking, the Gulf Stream was discovered in 1513 by Ponce de Leon, a Spanish explorer, but who knows how many others had already found and made use of it? Christopher Columbus had noted as early as 1492 the presence of a current in that region. Other navigators attempted to understand and describe it, but the first scientist to make a true study of it is unquestionably Benjamin Franklin, that universal and encyclopedic American, printer, revolutionary, pacifist, politician, ambassador, and above all, remarkable man of science. In 1769, shortly before the American War of Independence, Franklin was Postmaster General of the American colonies. At that time the Board of Customs in Boston was complaining to the honorable lords in London that official mail usually took two weeks longer to make the crossing from Europe than did private fishermen. (A weighty argument against nationalization of services and one that has been brought up often since.) There was nothing but silence from London, and Franklin decided to study the problem himself; in the course of a cruise to Nantucket, he happened to mention it to a sea captain. The latter told Franklin that when he was chasing whales he avoided getting into the "current" since it would carry him rapidly toward the east. He said that on many occasions he had encountered official English vessels in the current and had advised them to get out of it to save time. But, said he, these government captains were too knowledgeable, too sophisticated, to listen to the advice of American fishermen.

There was an echo of this incident two centuries later. Cruising in the Gulf Stream off Bermuda in a small motorboat, the late Rachel Carson, a writer who specialized in studies of the sea and ecological problems, encountered a heavy British tanker making its way southward against the Gulf Stream. She approached the vessel and through a megaphone requested that the captain come to the bridge. Much astonished, and figuring that he was dealing with a boating tourist who had lost her way, the officer of the watch called upon the captain, who asked Mrs. Carson what he could do for her. He could hardly believe his ears, almost strangled on his cigar, when he heard himself addressed as follows:

"Captain, you're on the wrong course. Get out of the Gulf Stream and go five miles farther west. You will gain more than three hours' time from here to Miami!"

No doubt by the time he recovered from his astonishment he was already in Miami, but one day he would understand that Rachel Carson was right, just as Franklin had been. As a general rule, cargo vessels today try in every way to take account of ocean currents, but there are still many captains who systematically ignore them.

13

THE PX-15

Once Grumman accepted the project in general terms, the next two important steps were to define the operation more precisely and to get on with the construction of the new mesoscaph. I mainly prepared the program, in collaboration with Walter Scott and his co-workers Walter Muench, Lee Geyer, Ray Munz, and Edmond Rabut.

My basic idea was to drift in the Gulf Stream for a month, in a submarine big enough for an operating crew and a scientific team to live in. The word "drift" calls for some explaining. Long drifting voyages had already been accomplished successfully on the surface; Thor Heyerdahl's celebrated voyage on the *Kon-Tiki* in 1947 is still well remembered. But a drifting expedition under water had never been thought of. There is a very simple reason for this: most submarine vessels, including most military submarines, are not made to float freely at a specific depth. Their hulls are more compressible than water, and this means that if they attempt to bring themselves into equilibrium the least tendency to descend will carry them into water that is slightly denser, but not sufficiently so, to compensate for the contraction of their hulls and the increase of specific weight; the relative density of the submarine will increase and the vessel will continue to descend. In the same way, if the submarine has a tendency to rise, it will come into water not sufficiently less dense to compensate for the relative loss of weight due to the expansion of the hull, and the vessel will continue to ascend. Consequently, a conventional submarine in homogeneous water cannot retain its depth unless it advances, stabilizing itself by its ailerons or its diving planes, or else by constantly changing its weight, taking on water to descend and expelling it to rise. Thus it remains stable to a certain degree, but these operations, which can be controlled by an automatic pilot, are costly in energy and moreover make noise, preventing many acoustical measurements. Therefore the solution lies in a much more rigid hull which will definitely be less compressible than the water it is in, and because of that fact will provide the stability necessary for a prolonged drift at a moderate depth. When such a submarine descends slightly, it will grow lighter in relation to the water, and the descent will have a tendency to stop by itself; if it rises slightly, it will grow heavier, or, to be exact, its specific gravity will increase in relation to the water and the ascent will stop by itself. But of course a heavier hull again brings up the matter of buoyancy, which is the basic problem of all submarines.

If one examines the plans of a naval submarine one is amazed to see how much weight is devoted to accessories which play only a negative role: torpedoes, guns, mines, and so forth. Eliminating these machines gains two advantages for the builder of a civilian submarine: first, he can have a hull that is relatively heavier and also less compressible than water, the advantages of which have just been noted; second, with a more resistant hull, he can descend to a greater depth without altering the coefficient of safety. Also the civilian submarine can save weight on electronic equipment, which is becoming more and more important in military submarines.

In general terms, the desired characteristics, or specifications, looked something like this:

Operational depth, 2,000 feet (about 600 meters)

Compressibility, less than the water at depths to be explored

Safety coefficient, minimum of 2

Payload for scientific equipment, minimum of 2 tons

Inside equipment, for six persons for a period of six weeks

As a matter of fact, the desired specifications were very close to those of the *Auguste Piccard,* though interior arrangement would of course be totally different.

I had even suggested to Grumman that the company buy the *Auguste Piccard,* which had been for sale since 1964, and negotiations had been started but came to nothing. From the beginning of our collaboration, the Grumman engineers had told me that they wanted to get all the necessary experience and know-how. If they bought a submarine they would learn how to adapt and modernize it and perhaps how to operate it, but they would not have gone through the problems of building it. As for myself, though I remained sentimentally attached to the first mesoscaph, I was delighted with the idea of building another, this time under ideal conditions.

We decided to make use of methods and techniques employed for the first mesoscaph wherever possible, but of course to improve on them everywhere we could.

Let us examine some of the specifications compared with those of the mesoscaph *Auguste Piccard.*

In choosing the maximum depth for the first mesoscaph I had two reasons for settling on approximately 700 meters, with a safety factor of 2. First, as a tourist submarine in Lake Geneva, which has a maximum depth of 310 meters, the vessel would always have a safety factor above 4. On the other hand, after the Exposition I wanted the vessel to be able to descend into the sea to between 600 and 700 meters, the limit to which daylight penetrates, with a safety factor of 2. The calculation then indicated that we needed plates 38 millimeters thick of the selected quality of steel.

For the new mesoscaph, I took as a point of departure the fact that Giovanola had 35-millimeter plates in stock. This was after determining the general lines of the hull and before I had received the telephone call from Grumman. The reduction of some 8 percent in the thickness of the plates was compensated for through special tests and by the remarkable quality of the work on Giovanola's part. On the other hand, granting that stability under water would be crucial, I decided to increase the rigidity of the hull a little by augmenting the thickness of the flanges of the reinforcing rings; the compressibility of the hull was thus somewhat further reduced.

It was decided to fix the maximum depth at 610 meters. It may seem artificial and arbitrary to determine permissible depth with such precision, but it will become clear that this is an indispensable measure, especially taking into account the future collaboration with the United States Navy. On the other hand, since funds were now less limited, I was authorized to increase by a little the length of the hull, and we arrived at the following final dimensions:

Outside diameter, 3.15 meters (same as the *Auguste Piccard*)

Inside diameter, 3.08 meters

Length of the cylindrical section, 11.60 meters

Total outside length (not counting the reinforcement of the portholes) 14.75 meters

Total length overall, 14.82 meters

The calculations for a submarine hull are in themselves a very complex problem, especially for a cylindrical form with reinforcements. For the mathematicians of our time, electronic computers are invaluable. After our hull was calculated by conventional methods, one of my fellow workers, François Hemmer, assistant at the Swiss Federal Institute of Technology, gave the problem to a computer in Zürich. Thus we had both a verification of our basic figures and instantaneously an abundance of extremely valuable data, analyzed centimeter by centimeter, and we knew for each specific point the tension of the metal and its deformation as a function of the depth attained. We also gained our first insight into the perspicacity of these machines. In programming the problem, the operator struck a wrong key, indicating the capital letter O instead of the zero. The computer would not let him get away with that and threatened to stop work immediately unless its instructions were dictated more correctly.

"See here," it said in effect, "are you sure that's really a capital O and not a zero?" (Its "thought" was transmitted on a printed band by the words: "Incorrect spelling, 0 or O?") Covered with confusion, the operator made his excuses, struck the cipher, and the computer went back to work.

The problem of our hull was put to another computer in Bethpage, quite independently of our researches in Switzerland. The results were exactly the same, not because the two computers had been made by the same company but because the problem had been properly programmed. A general formula of greater scope had been established by the Grumman mathematicians, and thenceforth we could put to them any general problem whatever about a cylindrical submarine hull and the computer would instantly give the most precise answers concerning the thickness of the plates, the geometry of the reinforcements, and the compressibility of the hull, for any given depth of any type of steel or other material.

Though this hull was quite similar to that of the first mesoscaph, the general arrangement of the new submarine was altogether different.

The *Auguste Piccard* had been designed as a tourist submarine with the underlying idea of later conversion into a research vessel if conditions permitted. Therefore it needed a good rate of speed and consequently as efficient a hydrodynamic shape as possible, and the interior arrangement previously described. The tail, or rather the end of the tail, designed for us by the firm of Kort in Hamburg, was a great success. The speed of the mesoscaph *Auguste Piccard*—6.3 knots—was obtained by only 75 horsepower on the drive shaft. This was made possible principally by the way the streams of water were carried to the propeller and then expelled through the nozzle. This scheme, outlined by us and executed in detail by the specialists, nevertheless had a drawback, unimportant for the *Auguste Piccard* but crucial for a research submarine: no visibility was possible from the stern.

For the mesoscaph designed to drift in the Gulf Stream and later to operate and observe in conditions difficult to foresee, such a blind spot was highly undesirable. Many fish, as well as porpoises, swim around a vessel, particularly if it is at rest on the surface. It was fundamental to my purpose to have the possibility of observation extending all around the hull and consequently not to have a tail or to reduce the tail to a minimum. The plan accepted placed the four motors outside the hull, two forward (starboard and port) and two at the stern (also starboard and port). These motors placed in the open water would have an acceptable efficiency, though naturally less than that of the single motor of the *Auguste Piccard.* Since in addition the keel had a cross section much greater than that in the first mesoscaph, the speed of the new machine would obviously be less, but with four independent motors mounted in pods and able to be tilted around an axis of almost 130°, maneuverability would be incomparably superior.

Selection of the motors required a complete study in itself. Although the passage of the drive shaft through the hull of the *Auguste Piccard* had been completely satisfactory, I wanted to avoid having four similar penetrations in the hull of the new mesoscaph. Besides, four big motors of 25 horsepower inside the mesoscaph would have been extremely cumbersome and noisy. It was preferable to place the motors outside and have them operate under equipression, as it is called. Besides, a motor, like any other accessory that is placed outside the hull in direct contact with the water, is lighter by as much as the weight of the water displaced. In practice this gain was negligible, for the external protection of the motors, even though it was not designed to resist the water pressure, had to be taken into account in the weight budget. The motors would have been made uselessly heavy if they were placed in watertight cases to resist the exterior pressure.

There are several ways of making motors operate under equipression, even in salt water. The simplest way, inaugurated during 1946–48 on the first bathyscaph by Dr. Auguste Piccard, consists of immersing the motor in oil and keeping this oil always at the exterior pressure. This is accomplished by the use of a rubber container that can stretch under pressure and by compression send enough oil into the light casing of the motor to maintain the interior pressure at the same level as that outside. We adopted a similar system again for the *Trieste,* using trichlorethylene as insulating liquid with a density of 1.4—that is, heavier than water. Since the motor was vertical even a tight joint for the shaft was not needed. Actually, the use of trichlorethylene caused all sorts of problems, for this liquid is a relentless solvent that tends to erode anything it comes near.

Although our motors had operated at 11,000 meters depth aboard the *Trieste,* and although firms of specialists had perfected motors of this kind that run in oil, I wanted to use a different system of which I had heard talk for some years—one that had been developed particularly in Hamburg, Germany, by the Pleuger company.

Pleuger had specialized in motors and pumps for use in very deep wells, as well as "accessory steering devices"—that is, additional propellers placed laterally which enormously facilitate the maneuvering of ships operating at low speed, especially in harbors.

For this purpose Pleuger made use of motors of three-phase alternating current without brushes or mobile contacts of any sort, and with especially designed rubber bearings instead of conventional ball bearings.

This reminds me that ten years before, at the time when I was in Castellammare di Stabia working aboard the *Trieste,* I had been visited by a German engineer who wanted to see the type of motor we were using for the bathyscaph. Since he told me that his firm specialized in motors of this sort, I showed him the plans of our system, then the machines themselves, and gave him all the details that could be of interest to him. When he left, one of my Italian co-workers reproached me for being too generous.

"On the one hand, you never have enough means for your researches, and on the other you divulge to anyone who will listen all your best secrets! It's a mistake, signore!"

I replied, citing La Fontaine, that a good deed is never wasted and that perhaps someday I would be in need of that engineer's knowledge, a remark that I forthwith forgot.

What was my surprise, arriving in Hamburg one fine morning in 1965, to be received at Pleuger's by this same engineer, Franz Zacharias, who in his turn left no stone unturned, first to satisfy my theoretical interest and then to fill the order we gave him. The reception I had given Pleuger's representative in Italy is certainly one of the reasons for the excellent relations that we now have with that company.

However, Pleuger motors use alternating current, and since our batteries could supply only direct current, we had the problem of conversion. The idea of a conventional generator producing alternating current and run by batteries did not appeal to me. From the beginning I had been tempted by electronic inverters with thyristors, which I had formerly considered using for the *Trieste.* But in that case only 1 or 2 kilowatts were required; here there was need for 120 at least. Another problem was how we were to control the speed of the propellers?

We studied a whole series of possibilities: variable pitch propellers, batteries connected in series and in parallel, various couplings of the motors, and so on. But it turned out that AEG in Hamburg and Berlin was making inverters of variable frequency and voltage; the company had even recently built for use aboard a trawler a 60-kilowatt inverter of variable frequency about which the reports were excellent. Just what we needed!

I can't say that from the first everything went without a hitch. Far from it. Even with AEG entirely cooperative, the problems to be solved were considerable. The electronic world is indeed a world apart. In this specific case very few engineers were able to formulate the problem perfectly or to understand exactly the function of these machines, even if their elementary principle is relatively simple. We had to have many discussions to establish the exact characteristics of our inverters, their output under various conditions, even their weight and size, all of which we had to know in advance with absolute precision. Several of our people had to make frequent trips to Hamburg and Berlin where the machines had been developed. We also had to make many tests, and even after the inverters had been installed we needed several months to familiarize ourselves with them completely. On the other hand, one advantage of electronics is that you do not have to understand every detail in order to make a necessary repair; you simply change the different printed circuits successively until the machine works. In fact, we are now completely satisfied with the solution we adopted. But it took a big dose of perseverance and patience, as much on AEG's part as on ours.

To supply these motors, the floodlights, the life support system, and in general everything that moves, lights up, warms, or serves as an indicator, a great many storage batteries were required. In the balance sheet of weights, I therefore assigned to this an important place, even more important than for the first mesoscaph, in fact about 20 percent of the total weight. But to augment the capacity of these batteries I decided to place them in the keel outside the mesoscaph, subjected directly to the pressure of the sea. This had first been done aboard the first bathyscaph by Dr. Auguste Piccard with very good results. The advantages of exterior batteries aboard a submarine are many: you economize on weight, in this particular case more than 40 percent; you save space in the interior; and you prevent the diffusion of gas in the living quarters in the interior. Lead-acid batteries, as a matter of fact, *always* have a tendency to produce hydrogen and to some extent oxygen. Hydrogen, particularly when it mixes with oxygen and reaches a proportion of 3.9 percent in the air, becomes an explosive gas which has already been responsible for the destruction of quite a few submarines.

In the *Auguste Piccard* the batteries had been placed within the hull, but there the conditions were different in that the dives had been planned for one hour's duration, and a general system of ventilating the cabin had been worked out to function, and indeed did function, rigorously between dives. For the new mesoscaph, even though we would not be submerged during the charging of the batteries (it is usually at the end of this process that most of the gas is released), we did not want to take the risk of remaining for a month or more in the same compartment with 26 tons of lead-acid batteries.

Since the mesoscaph is designed to perform a series of successive dives, we had to adopt a system that was entirely automatic to let the gas escape whenever necessary without getting an expert to come and open or close a valve before or after each dive and so as to avoid the risk of having the oil, which serves as an insulating cushion between the acid and the sea water, drawn into the sea with the gas from the batteries at the end of a dive. A new automatic system had to be developed, with all that that implies of enormous interest, but risk as well.

The new battery (70 kilograms in the air, 40 kilograms under water, 2 volts, 1,000 ampere-hours) had each of its elements provided with a cover in the form of a reservoir. This permitted storage of a relatively large reserve of the electrolyte and the oil that insulated the battery from the sea. At the top of the cover a pipe of polyvinyl chloride was inserted, leading to a large gas collector, itself equipped with automatic valves in several locations, which allowed the gas to escape but did not allow the sea water to enter. The collector in turn is connected to a central reservoir of oil open to the sea, the purpose of which is to maintain a constant pressure between the interior of the whole system, including the batteries, and the outside; it must in particular supply the oil to refill the space freed by the compression of the gas from the batteries, especially at the beginning of the dive.

The problem of gas from the batteries had been studied with great care, for it played a direct role in the stability of the mesoscaph. Each liter of air or gas has a buoyancy at the surface of about 1 kilogram, which is reduced to 500 grams at 10 meters depth, to 250 grams at 20 meters, and so on. Accordingly, we arranged every section of the batteries in such a way that the gas could escape with as nearly equal ease as possible, especially after charging. The elements were placed horizontally in the keel; the separators between the lead plates were especially chosen; the supports for the plates were arranged in such a way as to guide the gas toward the outside. In short, with all these precautions the zone of instability for the mesoscaph (of which more later) was reduced to a few tens of meters beneath the surface—that is, to a zone where the differences in temperature would in any case be to our advantage. This study was made in very close collaboration between the makers of the batteries, Electrona of Boudry, under the special direction of engineer Eugene Singer, and Christian Blanc in my office. The hundreds of connections between the various elements of the battery and then the connections with the interior of the hull, the hundreds of meters of cable carrying hundreds of amperes at hundreds of volts, all in salt water and in the shadow of Murphy's Law ("If anything can go wrong it will go wrong!"), certainly posed headaches enough for the engineer, the technician, and the workmen. However, as will be seen later the system was excellent and functioned perfectly; nevertheless it will be further improved in detail in the future.

The hull, the motors, the batteries, and the inverters—the principal elements of the PX-15—had thus been defined. Construction could begin. These four key elements were ordered from the suppliers at approximately the same time, at the end of 1966 and the beginning of 1967.

Very soon I had to expand my working team. It will be remembered at the time of the first mesoscaph I had got together an excellent group, homogeneous with one negligible exception, and that this group had, with the exception of the exception, been fired by the Exposition. Unfortunately I had not been able to keep the whole group with me; new jobs had been found for the men we had had to let go. Only Erwin Aebersold, Christian Blanc, and Gérard Baechler remained in my office, and though we could go ahead, it was at a slower pace. With these three key members of my office I laid out the different projects which I proposed to Grumman. Once the agreement with the company had been reached, I could increase my personnel again and get together again a team which restored to their places many of my former co-workers, in particular Pugliese, up to that time an engineer with the Ateliers de Construction Mécanique de Vevey, who joined me as assistant director. Pugliese had already had a good deal of experience, notably in the construction and interior fittings of the second cabin of the *Trieste* and in building the first mesoscaph. He was therefore an excellent acquisition. For some time our center of activity remained at Lausanne in my laboratory, which also had been enlarged; but more and more often I had to go to Monthey to the Giovanola factory where the hull was about to take shape.

14

GIOVANOLA

In Monthey, too, I found again many former collaborators; it is not possible for me to name them all, but I was especially happy to work closely again with Carl Guby, chief engineer; Auguste Chevalley, former chief engineer, then retired but always ready to lend us a hand; and Charles Weilguny, whom I have already mentioned.

The whole atmosphere of the Giovanola works has a special quality, difficult to define—that permits the work to proceed in remarkably pleasant and favorable circumstances. First of all, the location: at the foot of the Dents du Midi and the Dents de Morcles, where the valley of the Rhone broadens again after passing through the narrows of St. Maurice; in open country, country that already holds signs of the mountains, with its pastures, its fields, its vineyards, its pure air that comes down from the Val d'Illiez or steals up from the banks of Lake Geneva some 20 kilometers downstream, sometimes a light mist that rises from the Rhone and drifts through the ancient poplars that line its shores; the good heat of summer, never torrid, never overwhelming, and the good dry cold of winter, never excessive either; all these conditions give extraordinary advantages to the Ateliers de Construction Mécanique at Monthey. But, favorable though the geographic and climatic conditions may be for stimulating creativity in the workers, they are not sufficient by themselves; something else is needed, and there lies one of the great virtues of the Giovanola family. For this is quite simply a family business; founded nearly a century ago by one of the grandfathers, always in control of the family (the stock has never been put on the market), it is directed now by Marc Giovanola, grandson of the founder and son of Joseph Giovanola who was the company's principal developer. With each generation the Giovanolas have become more numerous; as directors, managers, engineers, foremen, workmen, they are ubiquitous in the factory, have the most entries in the company telephone book, and are always the main members of the board of directors. But what constitutes their strength is not their numbers or the amount of stock they own; it is their character. At Monthey there is no Giovanola empire but a Giovanola family, with all that this implies about human relations.

When Marc Giovanola is on his daily rounds in the factory it is hard to tell whether he is the father, cousin, brother, or manager, but everyone knows that he is every man's friend. It is not the old paternalism that was as much extolled as it was criticized. It is not an intentional policy; it is not socialism in any sense of the word. No, it is none of these. Perhaps it is nothing but a sort of spontaneous humanism inherent in a family which was created to manage a company of this sort. There are many legends about the Giovanola family, and unquestionably some of them are true. Others, like most historical legends, have the advantage of allowing one to glimpse a personality even if the facts are not exactly right. And after all, what one thinks he knows about a historical figure is often more important than the actuality.

One day a Giovanola workman announced to his friends that he was going to be married. It so happened that this Bolomey, as we shall call him, was a decent fellow but not very bright. His comrades often played jokes on him and on this occasion they said to him: "Go to the Boss, tell him you're going to get married, and he'll give you the most complete and up-to-date equipment for your kitchen."

"You don't mean it?" said Bolomey.

"Why of course, don't hesitate. It's the custom here. For each new household a new kitchen range!"

And Bolomey approached the manager.

"Mister, I'm going to get married—"

"Ah, splendid," said the manager. "All my best wishes."

Bolomey, slightly embarrassed, twisted his cap in his hands. After a moment he added: "It's about the kitchen, Mister."

"The kitchen, what do you mean?"

"Well, you see, they told me that when you get married you're given a kitchen."

At once the manager understood what was up and decided to turn the tables on the jokers.

"Yes," he said. "That's right. Take this note and go to the store on La Grand' Rue, buy a refrigerator, a kitchen range, a set of casserole dishes, all the accessories. Have them send me the bill."

Bolomey's malicious friends were waiting for him to return.

"Well now?" they asked as he entered the shop.

But his contented air troubled them. They stopped laughing when they heard him say: "Yes, everything's all set, I can buy the things. All I have to do is send him the bill."

15

AMERICANS IN LAUSANNE

Grumman had assigned some of its engineers to Lausanne to follow the construction of the mesoscaph, in particular Don Terrana and Al Kuhn. It was principally they who helped superintend every step of the construction of the hull and all the accessories of the PX-15.

A Texan by adoption if not by birth, Don had some trouble at first getting used to Switzerland. The traffic, especially in the narrow winding streets; the beautiful rural scenes one would have enjoyed much more if one did not have to get through so many tiny villages; the general skill of the European driver (a skill that an American tends rather to consider nervousness or madness)—all these things upset him more than once. But like a good tourist he adjusted to his new life, went in for skiing and private flying, traveled a great deal in the country and was surprised to discover that Switzerland produces other things besides cheese and watches. He came to admire Swiss industry highly and had an optimistic view of its possibilities. We for our part admired the precision of his judgment, the accuracy of his criticisms, and the advantages of certain American methods.

There is, by the way, nothing astonishing or exceptional these days in this sort of collaboration between an American and a European company, even if the European company is three thousand times smaller. It is wrong to believe that Americans do not know Europe. They have, as a matter of fact, a perfectly good knowledge of it, but they do not assimilate it, just as we Europeans have difficulty in assimilating transatlantic culture and methods. But in the procedures that Grumman introduced with us, there was no difference except that between America and Switzerland: a profusion of Coca-Cola, American working hours, the predominance of money problems. There was also, and I might say especially, the fundamental difference between a big company taking a plunge from aviation into the undersea world and a small group of technicians who had merely decided among themselves to build a new mesoscaph and to prepare for a 1,500-mile drift in the Gulf Stream. For Grumman everything was new; even the concept of a submarine was strange to them. The American system of measurements was a stumbling block, much more obvious for undersea work than for any other. With us, particularly in reference to the ocean, the cubic meter instantly called to mind one ton, and, accustomed as we are to the metric system, in speech we often interchanged the terms ton and cubic meter. For the American first using the metric system, there is a complex problem of conversion. Rulers and tables were worn out calculating feet and centimeters, short tons and metric tons, not to mention Fahrenheit and centigrade degrees. One day when Don was talking to us about an angle of 30°, we asked him whether the degrees were centigrade or Fahrenheit. It took him some while to make an appropriate reply.

In America, in Great Britain, everywhere, people recognize that the metric system is incomparably better. It has even been said that the head start in space held by the Russians for several years was due partly to the time lost in the United States by the difficult system of measurements. The fact that America has caught up brilliantly and in its turn made a formidable advance does not invalidate this hypothesis; it simply proves that Americans decided to pay the price.

However, it was not easy for the American engineers to adjust, from one day to the next, to the metric system, and many other things must have struck them as strange.

When Grumman had accepted the idea of building the new mesoscaph in Switzerland instead of in the United States, it was because the company had admitted—it should be clear to all—the advantages inherent in our methods and circumstances. A small, flexible organization, where everyone contributed as much as he was able and not simply in his particular field, could do the same work much more cheaply than could a giant organization in which everyone is so specialized that he dares not lay a hand to anything outside his assigned domain. What was the surprise, the dismay, of my Grumman collaborators when they realized that our people often went directly from the drawing board to the telephone to order material, to the car to go in search of material, to the workshop to adjust it, to the assembly hall to put it aboard the mesoscaph. Reports from Grumman's representatives flew to Bethpage, piling up in alarmed cascades: "It's the man who cleans up at night who is in charge of making the electrical circuits. A draftsman spent several hours adjusting a valve. An engineer wasted half a day choosing a fuse from a display!"

Actually, it was the other way around. It was not the man who cleaned up at night who made electrical circuits, but rather the electrician who helped in the evening with the necessary general cleaning up of the workroom. If the engineer or the draftsman adjusted a valve or selected electrical supplies, it was because they were the ones who knew the problem and could find the best solution.

Things like these marked the major difference between Grumman and my office. On our part, we wished to economize and remain within the fixed budget (and we did not exceed it); also we wanted to lose no time, for a few weeks' delay might mean a postponement of the Gulf Stream mission by a whole year, so we took pains to reduce the paper work as far as possible within normal requirements. Grumman, on the other hand, insisted, particularly at the start, on applying in Switzerland all the methods that succeeded so well in America. In the end, we arrived at a happy compromise that satisfied Grumman and caused us no great loss of time and no budget excess.

16

TECHNOLOGICAL DEVELOPMENTS

The work went well. We made a trip to Linz, in Austria, accompanied by Don Terrana and Carl Guby, chief engineer of Giovanola, to decide on the quality of steel for the hull. We were pleasantly received by the Voesst Company and shown the steel works which had got its share of American bombs during World War Two and had nevertheless survived. Because rancor plays little part in the Austrian character, and perhaps even more because business is business whatever the frontiers or nationalities, Voesst made every effort to provide us with the right steel for this submarine destined for America. The company had also furnished the steel for the first mesoscaph.

The choice of steel was interesting. Since practically all steels have about the same coefficient of elasticity, it is evident that a high ultimate breaking point, other things being equal, will permit a submarine of a given weight and volume to penetrate deeper into the sea than will a relatively mild steel—that is, one of medium hardness. While bearing in mind the need both for a steel with a good elastic limit and one affording a good possibility of elongation before breaking, the builder is often tempted to choose "the best steel," the strongest. But the "strongest" is not always the best. Modern metallurgy has developed extraordinary steels, three or four times stronger than good conventional steels. In the laboratory even greater advances have been made, but these steels are still not thoroughly understood; they are often brittle and generally difficult to weld. These are probably temporary disadvantages, and if one extrapolates the advances since the end of World War Two there is every reason to think that decided, if not definitive, progress will be made in the very near future. In our case, I frankly preferred, for a submarine that would make numerous dives in sometimes difficult and often varying conditions, to make use of a good conventional steel, of known quality, promising no surprises even though no sensational performance, rather than a steel with a very high breaking point but dubious as to its resistance where welded or where thermal treatment might not have been carried out uniformly. This can be summed up very simply: better to dive to 600 meters with complete safety than to 1,000 or 1,500 meters under constant dread of some irreparable accident.

American industry has developed some remarkable steel often used for submarines; the HY-80 is typical. Voesst in Austria had developed a similar steel, the Aldur 55, which I had chosen before for the first mesoscaph. Its yield point and breaking point are very close to those of the American HY-80, although it does not have the same resistance to shock. Shock resistance is an important quality in a military submarine, which has to withstand the violence of depth charges, but it is less important in a research submarine, which always runs at low speed and from the surface receives nothing but aid and comfort, not depth charges, mines, or other engines of death. Grumman readily accepted the Aldur tempered steel for the cylindrical part of the hull. But it took us some time to agree about the choice of steel for the two hemispheres. Aldur 55 cannot be forged or pressed without a reduction in its elastic limit through thermal treatment that weakens it. Although in a general way the hemisphere is less affected by the pressure than the cylinder is, I nevertheless wanted to have something better adapted to our needs for the PX-15. Krupp offered me an excellent steel, relatively mild, the Welmonil, which had practically the same elastic limit as the Aldur but which, not being tempered, could be forged or pressed with ease. Krupp had constructed the second cabin of the *Trieste* during 1958-59, the one that descended to 11,000 meters in January 1960, and I looked forward with pleasure to working again with that firm.

On the other hand, since the first mesoscaph had proved so safe, Grumman wished as few changes as possible and by telephone and letter kept insisting on having Aldur 55 for the hemispheres as well. I had to fly to New York to explain in detail my reasons for preferring the Welmonil. Finally understanding our view—that is, Giovanola's and mine, Grumman accepted the Welmonil and was well satisfied with it. The American Bureau of Shipping, which was to follow construction from beginning to end and give us a certificate of "good behavior" before starting the Gulf Stream expedition, was particularly interested in this steel and its welding possibilities. Although Welmonil was slightly less resistant to rupture than Aldur, so good were its characteristics under welding that it would almost have served for the cylindrical part of the hull without affecting the depth attainable.

The sheet steel began to arrive in Monthey in January 1967, and immediately after a final, extremely rigorous check of its quality, Giovanola could begin the first phase of the operation: rolling the sheets and forming the shell plates. The delicate operation of inserting the stiffeners in the shell plates was easier and better than what had been done in the case of the first mesoscaph. Instead of introducing the rings one by one into the cylindrical sections, these sections, as they came from the furnace, heated and therefore slightly expanded, were placed directly around the rings which were held in place by a temporary structure. The operation saved time, and was simpler and more precise.

As work on the hull advanced and the material to be mounted piled up, the center of activity gradually shifted from Lausanne to Monthey; car trips became more and more frequent, and once again we had to make up our minds to leave Lausanne and settle in Monthey. Giovanola prepared a large and spacious "workroom barracks" equipped with running water, heat, and electricity, which we immediately named the Gulf Stream Châlet. The two principal sections of the hull were to be set up in a new hall which Giovanola had been holding in reserve but now made ready for us, a large, bright, impeccably clean, and well-heated place. But before the hull could be installed there, another important operation had to be performed: the machining of the two main joints, which could not be done until after the thermal treatment.

One important characteristic of the hull of the PX-15 deserves mention here. When I had proposed this submarine to Grumman, I was immediately asked if I planned to include an airlock through which divers could enter and leave the hull. Though I knew how useful this device could be in many cases, I did not intend to provide the mesoscaph with one, for I thought the expense, complications, and inevitable delays could not be justified for a submarine intended primarily for explorations of long duration, such as the Gulf Stream mission. Grumman saw no immediate use for it either but nevertheless wanted to keep open the possibility of installing such an airlock later on.

Then there was another related problem. Like all welded constructions, the hull of the PX-15 had to be submitted so far as possible to a thermal treatment designed to reduce what are called the welding tensions. What a problem this had been with the first mesoscaph! For the PX-15, whose hull was a little shorter, we had two possible solutions: either enlarge the Giovanola furnace—and we had a firm offer for this project—or build the hull in two sections and join them with bolts. Thus, either way, all the welding would be annealed, something that the first mesoscaph had lacked. Each part of the hull could be provided with a machined flange that could easily be made as strong as all the rest of the hull. I had already had this idea, as has been seen, for the *Auguste Piccard,* but circumstances prevented me from making use of it.

Now, envisioning a flange that could be bolted and unbolted, we could also foresee the later insertion of a diving chamber with an airlock in it. The idea satisfied everyone, and we passed on to the blueprints for the flange and preparations for making it.

The principal technical problem of a flange of this sort is the precision that has to be obtained in machining the surfaces. We intended to use an O ring—that is, a toric joint inserted in a groove and rising only slightly above its edge in repose. This kind of joint is perfect if one has a good "metal-to-metal" contact when the rubber ring is slightly compressed. But to machine a flange 3.15 meters in diameter poses a difficult problem. Only highly specialized firms own large lathes or giant-sized milling machines capable of this kind of work. The one firm closest to Monthey so equipped was the Ateliers de Construction Mécanique de Vevey, which had an important part in constructing the *Auguste Piccard.* Vevey was willing to undertake the work and to carry it out on the new universal milling machine manufactured by Innocenti, a machine of large dimensions which the company had just acquired. Vevey was also to machine, with equal precision, the frames for the two access doors, the doors themselves, and the base of the four "feet" of the mesoscaph surrounding the keel.

The access doors to the interior of the mesoscaph also require some comment. When I sent the blueprints for these to Bethpage they were very carefully studied by specialists of the American Bureau of Shipping (ABS) and of the U.S. Navy. I awaited with curiosity their comments on this structure, which I myself had investigated with much care. Many pressure tests had been conducted in the laboratory, and models had been submitted to the merciless test of the pressure gauge which permits an exact determination of tensions in this sort of structure. I had taken every precaution to see that the frames reinforcing the doors, as well as the portholes, should be strong enough, but that it should not be *too* strong. In fact, a point exaggeratedly reinforced can lead to a premature rupture of the hull, for it prevents a uniform contraction under pressure. The assembly seemed light, it is true, but calculation and experiment showed that the door would probably hold. However, this design was so different from the conventional doors of submarines that the ABS wished to know more before "approving" it. Once more I had to fly to New York and spend several hours with the Grumman engineers and the ABS, foreseeing all the possibilities and explaining why and how I had been led to this conception. Finally the design was accepted without change. More than a year later, Matthew Letich, one of the chief engineers of the ABS with whom I had examined the problem, went on a 2,000-foot dive at the time when the mesoscaph was to be "accepted." He was then able to determine, making use of pressure gauges, that the deformations and tensions in the door and its framework were exactly what had been foreseen. He came back radiant from that dive. The collaboration with the ABS was open, cordial, and fruitful throughout the whole course of our construction.

When all the welding had been completed, the thermal treatment finished, and the hull given its first coats of paint, the two sections of the hull were sent separately from Monthey to Vevey. This transport was, to say the least, cumbersome, and it was sure to attract attention. We had to choose an hour when traffic would be relatively light and take little-used routes where possible. The highway police accompanied our convoy, and in order to cross one bridge over the Rhone it was even necessary to unload the cargo, send the heavy hauling truck through first, and then, by means of a cable draw the trailer with the hull over the bridge. Everything went off well; the machining was done in excellent style; the effective tightness of the doors and of the flange never gave us the slightest anxiety.

In the winter of 1967, after all the large elements —the AEG converters, the big aluminum panels, and the berths—had been put in place, a big step forward was accomplished in the closing of the hull. We brought the two halves together and bolted them. From that moment the mesoscaph began to take on its final form. It was then raised, so that the keel could be put in place.

This keel was more than an ordinary keel. It played, of course, a decisive role in stabilizing the mesoscaph but it had, as already noted, the additional function of housing the batteries. Made up of a multitude of cells, each containing a 2-kilowatt-hour element, it was constructed of steel and plastic as light as possible but also extremely strong. It had given a great deal of trouble to its builder Christian Blanc and to the supplier, the Eggli Company of Renens, but when everything was over, we could be proud of the result.

Grumman had undertaken to procure for us directly from America the water ballast tanks for the PX-15, as well as the so-called false ballasts. These were constructions of plastic, fiber glass, and polyester, built by a company that specialized in this sort of equipment for nuclear submarines. Thus we could expect specialized and expert work.

These were, you will recall, relatively light tanks under practically equal pressure with the sea, which supported the submarine on the surface and gave it good stability. Empty—that is to say, filled with air—

these ballast tanks gave the mesoscaph a buoyancy of 14 tons. To dive, it was only necessary to open the diving valves, then air could escape upward while water rushed in through permanent openings in the bottom of the ballast tanks. We gave the name false ballast to the plastic sheets that formed a transition between ballast tanks and the hull at the bow and the stern of the mesoscaph.

When these parts were finished, they weighed altogether close to 3 tons. Grumman nevertheless sent them to us by air. It was one of our first big surprises, though not the last. While we on our side were doing everything possible to economize, if not to the last centime at least to the last franc, Grumman did not hesitate to spend thousands of dollars to expedite by air in heavy cases these enormous ballast tanks with their mounting templates. So much the better, after all, if Grumman could afford it on its own budget, also very carefully figured in advance. Benjamin Franklin's saying, "Time is money," applied here in simple truth. Transportation by ship would have taken at least two months, and a delay of two months would probably have postponed the Gulf Stream mission for almost a year.

Despite the differences in techniques, norms, systems of weights and measures, and language that separated Grumman from Switzerland, these ballast tanks fitted on the hull surprisingly well. Only a minimum of adjustment was necessary, and another important step had been taken: the PX-15 was now visibly growing larger and plans were under way for its transport to America.

17

THE ROLE OF THE ABS

Every step had to be taken with exhaustive care, accompanied by tests. One of the people who became popular in Monthey was Renaldo Faresi, the superintendent attached to the Genoa office of the American Bureau of Shipping, who was charged with the inspection of our building operation.

It is perhaps important to make clearer the role of the ABS in the construction of the PX-15. As in every field in process of development—locomotives, for example, or automobiles, or airplanes—there was an era of absolute freedom for the builders of civilian submarines. It was this freedom that allowed the fabulous development of deep-sea submarines in the space of a single generation. William Beebe descended to 900 meters in 1934, suspended by a cable; the *Trieste,* a completely free-moving submarine, achieved over 3,000 meters in 1953 and nearly 11,000 in 1960. Such advances would have been absolutely impossible if there had been rigorous rules similar to those established years before in other fields, requiring a strict accounting every time one tightened a bolt or bought a voltmeter.

However, the very success of these machines for the peaceful exploration of the oceans gave rise, especially in America, to a craze for building similar or derivative devices. Consider the history of the first mesoscaph in Switzerland: civilian submarines were starting to be a means of locomotion not simply for private use of the builders but of guests, paying or otherwise, using the vehicle for their own pleasure or their own research. The submarine became a means of public service in a category similar to that of a bus or a taxicab. The responsibility of the builder was suddenly changed: on his own, he was free to make what trials he thought interesting, but as soon as he had passengers aboard, and especially passengers who had not been educated to understand the structure itself and the possible risks involved, a neutral agency of control had to intervene. This was why I had proposed that a specialist in the subject, Benvenuto Loser of Trieste, should follow the construction of the *Auguste Piccard.*

There are several official organizations of this sort. The best known are the American Bureau of Shipping (ABS) of New York, mentioned earlier, and Lloyd's of London, which follow and "certify" the building of ordinary ships. On the other hand, most national navies have their own internal bureaus, for it would obviously be undesirable for them to spread out their secrets for all to see. No such organization existed that was qualified by experience to follow the construction of a civilian submarine and then grant a certificate of navigability.

Grumman planned to apply to the U.S. Navy for this certificate. Coming from this source it would be recognized everywhere, needless to say, and would also facilitate the collaboration that was beginning between Bethpage and those circles of the Navy interested in submarine research. For a variety of reasons, principally because the vessel was being built abroad, this turned out not to be possible. However, we were at once put in touch with the ABS, which not only agreed on excellent terms to assist us in the mesoscaph's construction and to oversee practically every phase of it but even decided, as the building of the PX-15 advanced, to establish basic rules to follow in certifying other nonmilitary submarines. A committee was formed, including engineeers from various companies involved in the construction of submarines, Grumman among them, myself, and specialists from the U.S. Navy and the U.S. Coast Guard. And when the PX-15 was christened, the ABS published a manual setting forth the standards to be followed in the building of research submarines. This was not a "build-your-own-submarine" kind of thing but a detailed work with obligatory standards to which exceptions would be allowed only for specific reasons accepted in advance by the ABS and, if necessary, by the committee.

If this assemblage of rules ended the era of complete freedom, in America at least, it also marked the beginning of additional safety measures surrounding the whole technique, which had tempted many amateurs without sufficient knowledge or experience.

In practice, as various examples have shown, the aid and support of the ABS were of enormous value to us. True, they had no submarine specialists, but a submarine, like an airplane or a rocket, is nothing but a large assembly of relatively simple parts, depending essentially on metallic, mechanical, or electrical construction. The problems inherent in the submarine—such as stability, floatability, alteration of weight in the course of a dive—did not directly interest the ABS. The certificate was simply to guarantee that the mesoscaph had been built according to the rules and standards laid down, that steels were of required quality, that welding had been done competently and also had been checked by technicians, and that the techniques used had been of the highest order—in short, that the vessel was a sound structure. Faresi of the ABS came frequently to Monthey. His advice, comments, and always constructive criticism contributed substantially to the success of our enterprise.

Later on when it was decided to take one or two oceanographers from the U.S. Navy on the Gulf Stream Drift Mission, a new problem arose: the Navy no longer authorized its observers, oceanographers, or engineers to dive in vessels which it had not checked out. Gone were the times when the Office of Naval Research in Washington could send a whole group of scientists to Castellammare di Stabia on a dive program in the Mediterranean that was supervised by a Swiss. But with the PX-15 approved by the ABS, official sanction by the Navy was easily obtained. All the tests made by ABS were accepted at once by the Navy, and very few others were demanded; aside from the matter of form, it was a question of becoming acquainted with the vessel rather than examining it afresh. Submarine technique is still too young for routine to have become an intransigent and absolute mistress; common sense still reigns, if not by divine right, at least with enough flexibility so that the way is open for discussion.

18

DEPARTURE FROM EUROPE

During the winter of 1967-68 the outfitting of the PX-15 was completed at a good rate. I still had to make several trips to Bethpage, either to urge my views and obtain Grumman's agreement on some point or other, or simply to coordinate the work. Although it had long been decided that the mesoscaph should be built in Switzerland and later transported to America, the program after it reached the other side of the Atlantic was not yet definite.

Originally the first trials were intended to be made under my responsibility in Lake Geneva. I would have enjoyed this phase very much, for it is, in fact, one of the most fascinating in this kind of undertaking and one that had largely eluded me with the first mesoscaph. I would also have enjoyed being able to hand over to Grumman a finished submarine, in good working order and already tested, if not at its maximum depth (the lake is only 310 meters deep), at least deep enough to guarantee proper functioning.

Bethpage, however, was growing restive. Over there many of the directors, engineers and technicians, draftsmen and secretaries as well, heard talk about the mesoscaph but never saw anything definite except photographs, "which don't prove a great deal," and reports which were perhaps a part of an illusion. Did the mesoscaph really exist? Was it possible to construct a submarine in Switzerland—in the middle of the mountains? Hadn't the initial decision been a monstrous mistake? Some of the engineers came to Switzerland and assured themselves that the mesoscaph actually did exist and that the construction was proceeding at a rapid pace, but when they returned to Bethpage, they too met a certain skepticism.

There was only one way to establish credibility: bring the PX-15 to America and finish it there. Naturally enough, I was strongly opposed because it would be costly in money and time. I crossed the Atlantic once more and reached a new agreement with Grumman: the mesoscaph would be "practically" finished in Switzerland; then it would be quickly transported to America and all the trials made there. No doubt this solution had advantages for Grumman: whole regiments of technicians, who at the moment were twiddling their thumbs, could be turned loose to contribute a transistor, a bit of wire, a bolt, or a stroke of the paint brush to the object of their desire.

Grumman had also recently acquired certain retired naval officers who looked—why not say so?—most unfavorably upon the submarine's being first tested in fresh water, instead of what it was built for—salt water. Fresh water? Come now, did anyone really know whether fresh water is really wet? In any case, a test in fresh water would not prove a thing. As a matter of fact, fresh water must not be confused with distilled water; Lake Geneva, alas, is so polluted, so full of industrial wastes and acids that I imagine the conductivity is not much lower than that of ocean water.

The influence of retired naval officers in many large American companies is so extensive that it has even created a serious problem for the government. The education of naval officers may cost the government hundreds of thousands of dollars. If they begin their career at eighteen—the average age—they can retire at thirty-eight, and when they leave the Navy, whatever their rank, which is often that of captain or even admiral, they still have a whole future ahead of them. These officers have learned about the world, their country, and its industry through the psychology of the Navy, a psychology with which as a matter of course they have been stuffed. They are familiar with the Navy's past ventures and future projects, both long-range and short-range, with contracts in preparation, future appointments of officers, with petty gossip as well as great ideas. Thus they have an extremely high market value to industry and are literally snapped up when they retire. They can easily place with a former comrade a sizable order for aircraft, torpedoes, computers. Thus the acquisition of a retired naval officer by a company is justified at any price and is worth every consideration. The practice went so far that Congress had to do something about it, and for some years there has been a rule that a retired senior officer engaged by industry is not allowed to negotiate a contract with any government agency for several years after his retirement. Nevertheless, the prestige of these retired officers has remained great, though in our case their role was a bit different. It was their past experience, incontestably valuable, that tipped the scale in favor of making the trials in the United States and not in Switzerland, in ocean water rather than in Lake Geneva.

The work at Monthey continued without interruption. March 1, 1968, had been fixed for finishing the assembly. After this date, we had three weeks to disassemble the mesoscaph and make it transportable by rail and by boat. I insisted—all of us Swiss insisted—that the PX-15 should be presented complete in Switzerland, practically ready for launching.

All went well. There were almost fifty of us, technicians, workers, engineers, all working with the same will to have the mesoscaph ready on March 1.

The problem of transportation had been worked out in general terms at the beginning of the undertaking, but now the details had to be settled. The organization of this was entrusted to the Danzas Company that had arranged the transportation of Dr. Auguste Piccard's first stratospheric cabin from Brussels to Augsburg back in 1931 and the first cabin of the *Trieste* from Terni to Castellammare di Stabia in 1953. The mesoscaph would obviously have to be loaded partly disassembled on a cargo ship for the Atlantic crossing, but there were various possibilities open for the trip from Switzerland to the sea. I favored transporting the mesoscaph by train from Switzerland to Belgium and shipping it from Antwerp. I had known the port of Antwerp since 1948, when the FNRS-2—the father, in a manner of speaking, if not of all research submarines at least of many of them, including the bathyscaph and the mesoscaph series—left that port for the Cape Verde Islands off the coast of Dakar. I knew that if we went through Antwerp again we would be shown all the friendship and collaboration that the Belgians know how to bestow on the accomplishment of such a project. Fortunately, of all the possible routes—Monthey-Marseilles, Monthey-Le Havre, Monthey-Rotterdam, Monthey-Hamburg—Monthey-Antwerp proved to be the cheapest and fastest; this argument obviously was more important to Grumman than any historical or sentimental associations.

The land trip was to be by train, because, given the dimensions of the mesoscaph, it was much easier than by road. I had arranged from the start that the ballast tanks should be detachable, and the accessories welded to the outside of the hull were within the limits of size set by the international railroads. Moreover, the Swiss Railroad (CFF) assured us of their full and active cooperation. It turned out that several of their specialists, in particular René Aguet of Lausanne, had already shared the responsibility of transporting the first mesoscaph from Monthey to Le Bouveret. Thus in our eyes he was already an old submariner, and his experience was of great value to us.

On April 4, 1968, in the presence of a crowd of friends, fellow workers, and guests, the great door of the mounting hall was opened, and, drawn by a diesel locomotive made by Giovanola, the mesoscaph slowly emerged with a long blast from its powerful siren and was greeted, on a day that had been mostly rainy, by a brilliant and promising ray of sunshine. That same day it left the factory at Monthey, and Giovanola handed us an exit permit for "one mesoscaph (1)." It will be recalled that this was the second such document I had received.

In its trip across Europe by train, the mesoscaph passed many historic sites, going down the entire Rhine valley with its castles, crossing the famous bridge at Cologne below the great cathedral. The most striking contrast between two epochs was marked by a courteous gesture on the part of the CFF and in particular the station master at St. Maurice, who hastened the departure of the convoy by a few minutes so as to allow it to pause briefly beneath the Castle of Chillon on the shore of Lake Geneva. More than six centuries separated the two structures; the dukes of Savoy had never dreamed that the windows they put in their thick walls would one day witness the passage of a machine destined to explore a realm which to them was even more mysterious than the moon. A few minutes, time to take a photograph. . . .

Historical associations were numerous in Antwerp. The first bathyscaph, the FNRS-2, had been assembled there in the workshop of the Mercantile Marine Engineering Company. (The first FNRS was Dr. Auguste Piccard's stratospheric balloon.) This workshop again did us a service in 1968: we could test under rigorous inspection the suspension system that was to be used in loading the mesoscaph aboard the freighter. The foreman charged with making the cradle on which the PX-15 would rest during its Atlantic crossing turned out to be the same man who made the cradle for the FNRS-2 in 1948.

Finally, and most important of all, we were honored by a royal visit. Like his grandfather King Albert at the time of the preparations for the stratospheric ascents of Professor Piccard, like his father King Leopold before a later balloon ascent, and like his grandmother Queen Elizabeth when the FNRS-2 departed for Dakar, King Baudouin visited us and emphasized by his presence the interest he and his family had in scientific research. It is of great importance, particularly in a small country, that the chief of state should take a lively, personal interest in the great problems of the day. This is more than a tradition in Belgium; it is natural and spontaneous and helps explain why Belgium remains in the first rank of industrial and scientific nations in Western Europe.

In Antwerp, the mesoscaph was taken in charge by the Armement Deppe, an important shipping company whose freighters made regular trips from Belgium to Florida. Theoretically no Belgian line put in at Palm Beach, our port of destination; Belgian ships touched either at Fort Lauderdale or Miami, 80 and 100 kilometers, respectively, south of the port of Palm Beach. Though the distance was not great, transport by road would have been complicated and we wished to avoid it. The Armement Deppe made an exception in our favor and agreed to deliver the mesoscaph to Palm Beach on the M.S. *Anvers.*

The loading of an object of such weight—100 tons, even stripped of all its accessories, keel feet, ballast tanks, conning tower, rudder, variable ballast tanks, pipes, electric cables—is very impressive. Of course the crane hooks had been tested; the cranes themselves were constantly under precise and very exacting supervision. The workmen were highly qualified specialists. Yet there is always a certain risk, let us say a greater risk when a mesoscaph is in the air than when it is in the water. I have spent months, even years, in workyards, and I have seen a number of accidents, happily no grave ones; but I remember a day when a whole railroad car fell 10 meters because of a broken cable. A tested cable? Handled by qualified workmen? An accident is always possible; the first rule, moreover, is never to walk under a load. But on that day in Antwerp everything went well.

The cables were secured, the shackles fastened, the cotter pins inspected, and the PX-15 rose gently toward the gray but rainless sky. On every level of the freighter, from the main deck to the poop, those who were not engaged in the operation leaned out to look, some on the shrouds, some on handrails and gates, bending forward, sideways, backward, many with cameras extended, performing prodigies of equilibration in order to secure a pictorial record. Among the photographers and reporters present I recognized some who had attended the departure of the FNRS for the stratosphere in 1931 and the FNRS-2 for Africa in 1948. One was G. Champroux, an excellent photographer and faithful, devoted friend who had followed us on several expeditions.

The winches turned, the cranes groaned, great jets of compressed air and steam emerged, cables tightened, others relaxed, grew supple; voices crisp and precise, guttural voices of sailors from the north, issued brief orders that were instantly obeyed. With the gentleness and precision of a cat transporting one of her young by its nape, the crane deposited the mesoscaph on its new cradle. Well secured, it was ready to confront even a violent storm in the Atlantic if necessary.

After that, the accessories and the matériel had to be loaded; some 130 cases filling the ten freight cars that had accompanied the hull. The loading operation was accomplished peacefully, without problems, and with an almost miraculously easy routine. Whatever their form or weight, the cases as they disappeared into the enormous holds of the *Anvers* always found their exact places, as though an enormous jigsaw puzzle had been laid out in advance. What was our surprise to see that some of the matériel for the mesoscaph was loaded on top of cases of cheese clearly marked: "Real Gruyère from Holland, made in Finland." No wonder the American taste for cheese is somewhat different from ours.

With the departure of the mesoscaph, the European stage of our venture had come to an end.

19

SWISS IN FLORIDA

We had to reduce our team again, but in what different conditions from those in 1964! Many of our fellow workers had been engaged on a temporary basis, aware that they would not accompany the mesoscaph to America. They would have no trouble finding work in Switzerland, but in order to facilitate the reassembly of the mesoscaph in America I had persuaded Grumman to retain and take to Florida a basic group of those best acquainted with the mesoscaph. Thirteen of us—eleven Swiss, a Frenchman, and a Spaniard—set off for Palm Beach, arriving with families and baggage in successive waves, but almost all in time to see the PX-15 arrive aboard the *Anvers,* which despite some rough weather had had an excellent crossing. For all except me, this was the first visit in the United States.

The first time I was in Florida, in 1956, I found there, as everywhere else in America, a passion for the sea which at that time was equaled only by the ignorance about research submarines. Oceanography was studied from the surface, and soundings were taken by scientists perched on the rails of splendid sailing vessels. The necessity or even the possibility of sending observers into the depths had not then been recognized. Of course everyone knew about the dives of William Beebe, made during the 1930s, and everyone had heard about the bathyscaph which had conquered the Mediterranean. But in the eyes of most American scientists these were merely exceptional exploits which would not revolutionize methods more than a hundred years old. I have described in *Seven Miles Down* how I had found it necessary, aided by some American oceanographers including Robert Dietz, who understood the problem, the co-author of that book, to explain and re-explain the possibilities of the bathyscaph in general and of the *Trieste* in particular, to try to convince American circles of the possibilities and future of our methods. The efforts made at that time were supported, especially from 1957 on, by the Office of Naval Research in Washington. In 1957 the ONR had financed a series of *Trieste* dives off Capri and Ponza; then, convinced of the advantages of the bathyscaph, it bought the vessel, sent it to San Diego and financed the Guam expedition. In the course of that expedition, we descended to a depth of 11,000 meters. On January 23, 1960, the systematic support of the Navy and the incontestable value and accuracy of the data acquired by oceanographers using the *Trieste* brought increased recognition. Moreover, the economic opportunities that gradually developed—booming sales for scuba equipment, snorkels, and so forth; the increasingly frequent appearance of submarine films in theaters and on television; the sales of the stock of various companies interested in oceanography—all made the general public more and more aware of science vis-à-vis the sea. Two tragedies made obvious the usefulness of small research submarines. One was the loss of the submarine *Thresher*, the wreck of which was located and photographed by the *Trieste* in the course of a gigantic search operation. The other was the loss of a hydrogen bomb off Palomares, in Spain, which was located by the *Alvin*, a small Navy submarine under the supervision of Dr. Al Vine of Woods Hole Oceanographic Institution. In both cases direct inspection by men through portholes permitted an evaluation of the situation that no electronic or mechanical device could afford.

All this gradually lent confidence to industry: larger funds were allotted to the construction of small submarines, some specialized, others of more general utility. Indeed, it seems that years of delay were then more than made up. Supply exceeded demand; of the many small submarines theoretically in service only a very small number could bring any financial advantage to their owners. But this, let us hope, is a temporary situation, and probably this fleet (or at least its best members) will soon be in use full time.

Today, in Florida and all along the Atlantic coast a craze for things of the sea has seized the public. Everyone wants to get into oceanography, and the purchase of a snorkel may be the first step toward acquiring a Ph. D. in marine biology. The local authorities, the universities, the banks, and the churches, all support the movement.

Upon our arrival at Riviera Beach, Florida, just north of Palm Beach, in 1968, we were greeted on all sides with a friendliness that embraced the mesoscaph, the Gulf Stream project, and all of us who had come from so far away. Local authorities and companies gave us all the help we could use, and the work was rapidly resumed.

The mesoscaph was launched on July 25, 1968, and on August 21, christening it with water from the Seven Seas, we named it *Ben Franklin* in honor of the pioneer who first studied the Gulf Stream.

20

FIRST DIVE IN THE SEA

The first dive of the *Ben Franklin* was heroic. It involved a descent to the bottom of the harbor, but what precautions were taken, what responsibilities and powers were delegated on that day! The whole bottom of the harbor had been thoroughly explored by divers Erwin Aebersold, Gérard Baechler, Michel Page, and Marcel Dougoud; every shell, every discarded tire, every tree trunk had been seen and noted on a chart. After the harbor had been fine-combed, the precise location of the dive was studied and selected after studies by a committee. Every imaginable means of rescue had been thought of: emergency services, medical aid, firemen, Coast Guard, the most powerful lifting devices. With the capital commanded that day, not counting the

Ben Franklin itself, I could have built a new submarine. But what comfort, what a sense of safety, to be able to operate under such ideal conditions. Even the United States Navy, which is generous with money when it is doing research, had probably not expended as much on the *Trieste* when we made our descent at Guam, a thousand times deeper than we were about to go in Palm Beach. Everyone crowded around the mesoscaph, precise instructions in hand; the work had been so widely distributed that there was practically nothing for any one individual to do. When all the check lists had been marked off, the crew—there were six of us—went aboard. There, additional check lists had to be run through: testing the air pressure in the cylinders, the level of the oxygen, the presence of water, the emergency supplies, the working of the electric circuits, the functioning of the valves, and those thousand little nothings that can turn into catastrophe if they are not exactly the way they should be. Grumman was making its first dive; it had to be a success. The *Ben Franklin* had another characteristic important to Grumman: it was the largest machine the firm had ever built. Incidentally, it was the smallest one that I had ever built.

When all was ready, Walter Muench, director of the program, solemnly came aboard and asked us individually if we were willing to take part in this dive. With the voice of a doge of Venice when marrying the sea, each replied: "I am willing." Then Walt reminded us that this was to be a static dive, that we were simply going to rest on the bottom at a depth of 10 meters, that ropes would constantly safeguard us, that we would also be in telephonic communication. He asked if we were willing to take the risk of this dive, even though the inverters for the propulsion motors were not yet in final working order. It was only on a renewed affirmative reply from each of us that he gave us his blessing, remounted the ladder of the access port, and we were able to close the hatch.

The six aboard were Bill Rand, director of operations; Don Kazimir, skipper; Erwin Aebersold, chief pilot; Harold Dorr, co-pilot; John Greve, electrical engineer; and myself. Grumman wanted all the people responsible for the program to be present. Of course this was not a scientific dive but merely a technical trial.

The door panels closed; a cry resounded: "Fire, fire!" It was Don Kazimir, who wanted to start by having a practical exercise. There are a number of fire extinguishers aboard; Americans are always ready to combat fire with a maximum of technical efficiency. Everyone must be trained for this kind of emergency. At Kazimir's cry each of us took up the post assigned to him in advance, extinguishers were pulled from their brackets, and, since everything in the mesoscaph is metal and nothing would burn anyway, the extinguishers were put back in their places. Barely was this done than there came another cry of "Fire!" We fell upon the extinguishers once more and took up our positions. Instead of fire there was heat, 22° C., and great humidity, 83 percent. The oxygen was normal, 20.8 percent. No carbon dioxide was perceptible as yet.

Various additional tests were made before diving. The flood valves were opened, the weight of the mesoscaph was estimated, the ballast tanks were blown again, a little ballast was added to the bridge. The trim pumps that vary the attitude of the mesoscaph were tested. Operations had to be interrupted for a bit because we were obstructing the passage of a ship that had to leave port.

At 2:07 p.m. the four ballast valves were opened and the water rushed into the ballast tanks. Slowly, very slowly, the water came up—that is to say, little by little it covered the upper portholes, the hatches, the bridge. The temperature rose too. Now it was 24.6° C., with 89 percent humidity. The hatches, which, now just below the water, were in the most ticklish position in regard to watertightness (at greater depth the pressure itself helps to keep them watertight), did not admit a single drop of water. The hull penetrators for the electric and hydraulic lines were equally tight. It seemed that there would be little difficulty during the descent. The water was so calm and the mesoscaph so well stabilized (beyond the ordinary because of the ropes which secure it to the dock, to avoid our being swept away by the tide) that small quantities of air probably remained trapped in the ballast tanks. For several minutes the system of trim pumps continued to operate, and at 3:00 p.m. the very gradual descent began. At 3:05 the *Ben Franklin* touched the bottom of the harbor at a depth of about 12 meters. Once more there were certain tests to perform or parts to examine: joints, hatches, cable and pipe penetrators, and the great circular joint in the hull (the famous joint that Vevey machined for us); all was absolutely watertight, so probably the same would be true at a greater depth.

From 3:30 to 3:45 the airlock was tested. In Monthey we had built a device made from a pressure-resistant tube and two covers, also pressure-resistant, welded to the top of the hull which functioned as an airlock and permitted small objects *lighter than water* to emerge from the mesoscaph. One simply opens the lower cover, inserts the object—in practice a small sphere of Plexiglas or aluminum not more than 14 centimeters in diameter and containing, for example, a message—and recloses the cover, then with the aid of a small hydraulic pump the upper cover is opened, after, of course, filling the chamber with water; the light sphere rises, and those at the surface can recover it and remove the contents. This system had been constructed principally for use in the Gulf Stream. It is obviously a one-way system, but a return method had also been thought out: a message placed in a sphere heavier than water would be entrusted to a porpoise which could carry it directly to the little airlock of the mesoscaph.

It is known that porpoises are rated as second in intelligence to man. (What they think about this themselves is not yet known.) In California and Florida there are many training centers for porpoises. There is even a television program in which they are chief characters. With specialists at the Ocean World in Fort Lauderdale, Florida, I had discussed the possibility of training porpoises for this work. It is easy to imagine a police dog employed in activities of this sort ashore, but the porpoise is indisputably more "intelligent" than the dog. (Here too we need a better definition of the word.) The operation was almost decided upon with Grumman, but unfortunately we did not have time to carry it out. Also, it would have been relatively costly, for the training would have taken a long time, even not counting the danger of losing a porpoise if in the course of the operation it encountered another porpoise who might recall the taste for freedom, a more likely event if the other happened to be of the opposite sex.

On that particular day—it was still November 22,

1968—we had no porpoise and not even any special message to send to the surface. But we wanted to try out the airlock and send our remembrances to the general management of Grumman. So we dispatched a small bottle containing a letter addressed to William Zarkowsky, a Grumman vice-president who had followed in great detail the completion of the mesoscaph, and another to Walter Scott, the director of Grumman's Ocean System. These letters were to be stamped the same evening by the post office in Palm Beach. A diver, Gérard Baechler, watched the bottle emerge and ascend to the surface and took possession of it there.

At 4:15 p.m. we tried out the floodlights. They functioned well; the water was not really clear and the spectacle they revealed lacked interest.

The temperature did not rise—it remained at 24.6° C. The water at the bottom of the harbor is a little less warm than that at the surface, which compensates for the natural tendency toward elevation of temperature inside the hull. On the other hand, the humidity rose, now 94 percent. Obviously an effective system to deal with the humidity would have to be arranged. From the beginning I had been recommending the use of silica gel, which had worked very well in the bathyscaph.

That evening we were to proceed with a final test, the dumping of the emergency ballast. This ballast can be jettisoned in different ways: either gradually, by interrupting the electric current in the magnetic valves as I had done in the first mesoscaph, or by opening a hydraulic gate which, by bypassing the valves, empties the entire load at once. In case one of these gates should be blocked for some reason or other—by a shellfish, for example—a counterpressure could still be exerted which would force the gate to open. Finally, if one wished to block the iron shot for a long period, this could be done by magnetizing it, using a system like that which I had also installed aboard the *Auguste Piccard.* On this occasion we tried out the different magnetic systems first to get rid of a few kilograms of ballast, then at 4:35 p.m. we used the hydraulic system to make our ascent.

Before 4:36 we were at the surface; at 4:37 the ballast tanks had been emptied; at 4:38 the doors were opened. The first dive of the *Ben Franklin* was completed. Everything had proceeded without incident. We could now make preparations for the second dive.

There is no space here to tell about that in detail, nor to recount the many trials we made during the winter of 1968-69 and the spring of 1969. Grumman insisted on doing things very methodically and with that sumptuousness that was habitual with them, though not always in the way that I, as a European, would have recommended. Just the same, the result was good. Little by little each person became familiar with his new responsibilities. Many among the new personnel had to acquire experience of the sea, which they completely lacked and which the Grumman sailors, particularly Bill Rand, Don Kazimir, Harold Dorr, and Bruce Sorensen indefatigably inculcated.

On December 11 we began the first relatively long trial: an uninterrupted dive of three days' duration.

21

THE THREE-DAY DIVE

The great difference between the *Ben Franklin* and all other research submersibles is obviously its greater roominess and the arrangements permitting it to make dives lasting a month or more. It was essential at this point to proceed with one or more trials of sufficient length to establish an equilibrium in all those matters relating to the life-support system: essentially, the composition of the air (oxygen, carbon dioxide, moisture), interior temperature, average consumption of electricity, as well as to become accustomed to the food, to the hours of work, watch, and rest—in short, to the general ambience.

The dive lasting from December 11 to December 14, 1968, proceeded entirely according to plan and allowed us to make certain observations that were important for the rest of the program. The crew this time consisted of Don Kazimir, Erwin Aebersold, Harold Dorr, John Greve, Ray Davis, and myself. Davis, a microbiologist, was going to spend the three days microbe-hunting to try to find out where the microbes liked to hide, where they might come out into the open, and what had to be done to make them leave us in peace.

On December 11, at 10:45 a.m., we left the wharf to take up our position in the middle of the harbor basin. At 11:34 the hatches were closed. At 11:38 the flood valves were opened, and at 11:47 we were on the bottom. This time in the course of the three days we hoped, if not to make sensational submarine discoveries, at least to devote ourselves to some interesting outside observations. When we arrived at the bottom the visibility was rather poor and the few fish swimming around us were visible only at a distance of one or two meters. When the tide came in, the visibility would improve, thanks to the great quantity of clean water that the sea brings all the way into the harbor.

Since there were six of us, we took the watch two at a time for four hours. I was on the first watch with Harold Dorr, whose calm, confidence, and efficiency I have always liked.

Actually, there was not a great deal to do, the life-support system being almost entirely automatic, and I could devote many hours to looking through the portholes. However, there were a large number of regular observations to be made. In addition to the content of oxygen and carbon dioxide in the air, there was occasion to make sure that no toxic gas appeared in the atmosphere of the mesoscaph. One could easily imagine, for example, that if the insulation of the electric cables became heated for some reason or other, it might produce toxins. Generally speaking, on short dives these gases are not dangerous in small quantities, but the problem becomes completely different when one is shut up on board for several days or for weeks.

So we had a series of test tubes supplied by the Draeger Company of Lübeck, Germany, that permitted us to detect infinitesimal percentages of the principal "toxic gases" that might occur. The latter ranged from acetone to hydrazine, passing through trichlorethylene, toluene, trichloride of carbon, phosgene, and ozone —not to mention more than twenty-five other "deleterious" gases. Let me say at once that in the course of those three days only traces of acetone, methyl bromide, and olefins were detected.

The level of carbon dioxide was regulated without difficulty by panels of lithium hydroxide, which absorbs and stabilizes it. The panels could be changed at need. But it should be mentioned that the different instruments for measuring carbon dioxide gave somewhat contradictory readings. Fortunately carbon dioxide has the characteristic of producing headache long before it becomes really dangerous. Moreover, the panels of lithium hydroxide remain warm as long as they are active; when they cool off, it is time to change them.

The problem of humidity, on the other hand, presented certain difficulties. As I have mentioned, I had always recommended the use of silica gel, but the specialists were still searching for something with the advantages of this product and not its inconveniences, something essentially lighter and less cumbersome, for example. Since such a substance had not yet been found, and silica gel was not being used, the percentage of humidity increased very rapidly to the saturation point, producing condensation on the walls of the hull, and then a state of interior rainfall. It was not one of those diluvial rains characteristic of Florida, but enough water drops to blot all our notes, books, the papers on which we wrote or which we were studying. Moreover, everything quickly absorbed the humidity, in particular our clothes, the bedclothes and the covers, and at night or during the rest periods it sometimes became a real Chinese torture. The drops fell, intrusive, chilling, perfidious, without waking you up, for you knew what they were, but nevertheless without letting you sleep in peace.

At 3:15 p.m. visitors arrived. Gérard Baechler and Bruce Sorensen equipped with aqualungs, came to look in on us through the portholes. It was hard to recognize them, but I singled out Baechler by his equipment, which he brought from Switzerland. They communicated with us by gestures and with little cards on which they wrote that the mesoscaph had moved 3 meters since it landed on the bottom. This, moreover, was confirmed by the reference points which Erwin Aebersold has noted through the portholes. The temperature inside the mesoscaph increased slowly, whereas that of the water was constant (26° C.). It would be interesting to see in what range the difference of temperature between interior and exterior would become established.

In the course of the afternoon a handsome ray 60 to 70 centimeters in size visited us: some small "aquarium fish" arrived too. These were moonfish, angelfish, butterfly fish, and various other kinds of tropical fish.

Just before 6:30 we lit one and another of the floodlights: soon a cloud of little specks of plankton, apparently copepods, accumulated around the light. They were positively phototropic.

At 6:30 we had supper: I cannot say that the menu was appetizing—far from it; but this meal was symbolic. It was the first we had taken together aboard the mesoscaph. And, barring the humidity, our submarine was so comfortable, so friendly, the calm in which we found ourselves so beneficent after the agitation of the last days, that we had a wonderful memory of this occasion. About our round table each one opened a can of food, or dehydrated chicken "reconstituted" with water. For encouragement we had been told that this is the form of food prepared especially by NASA for the astronauts, and since everything that NASA does is perfect, this food should be excellent. We expected to see feathers sprout on the wings of our chicken, life return as after hibernation, and we pretended not to be disappointed when the pap which results from the absorption of water by the freeze-dried powder determinedly retains its aspect of pap. Bah! It would only last an evening, three days, or at the very most a month.

After supper I lay down for a few hours on my berth. To tell the truth, I could hardly sleep. Nevertheless the berth was comfortable, and best of all, it had a porthole, I could look into the sea above me. This evening there was not much to see, but I already imagined what the night would be like in the Gulf Stream. By interphone I asked Erwin Aebersold, who had the watch, to turn on one of the floodlights; pretty soon life began to swarm above me: the plankton was abundant, even in the water of the harbor, particularly murky at this time, for the tide was low. At midnight I resumed the watch with Harold Dorr. For half an hour I talked with Walter Muench on the telephone. He described at length the new Cabinet appointed by President Nixon. Walt is a philosopher. If he is at home with computers, he is nonetheless human. The news of the day which he described and remarked upon was more meaningful than if I had read it in the paper; he has a gift of presenting things in a lively and attractive fashion. A jet plane has been hijacked, rerouted to Cuba. Actually, if a countercurrent exists under the Gulf Stream, isn't there the risk that the *Ben Franklin* someday next year during its real mission, instead of drifting toward the north, might travel toward the south and be stranded on Castro's shores?

For this conversation we used the conventional telephone with a wire, but during the day we had already tried out the true submarine telephone, the "acoustic" phone that allows us to talk to the surface without wire, no matter what the depth.

During my watch, I could look out at leisure through the portholes. Though the water was turbid, numerous fish could be seen swimming around us. In particular I noted transparent alevins, very quick and about 1.5 centimeters long. A little later came schools of fish about 15 centimeters long, silvery gray in color, each with a handsome black spot on each side of the base of the tail; the tail itself is fringed with yellow. These are crevalle jacks, common around Florida and resembling small pompanos.

The percentage of oxygen in the submarine slowly rose: from 20.6 percent at noon it had risen now to 21.5 percent—1 percent in twelve hours. This means that the release of oxygen is unnecessarily great. The oxygen that we use is preserved aboard in liquid form—that is, lower than 180° C. below zero, in two cylinders that are perfectly insulated thermically. The flow can be regulated at will. At first we had fixed it at 3 liters per minute, corresponding, that is, to 0.5 liter per minute per person. At 4:00 a.m. I reduced it to 2.5 liters per minute.

Many fish seem to be attracted by light, but not all, at least not all in every case. It often seems that it is especially the plankton that are attracted and that they collect close to the lights. The fish arrive a little later, perhaps attracted more by the plankton, their food, than by the light itself.

At 2:22 a.m. I turned on one of the rear floodlights and saw a fine school of catfish, the fish that owe their name to their whiskers. As soon as they perceived the light they disappeared; on the other hand, they were not disturbed by the light, obviously less bright, that came from the interior of the mesoscaph through the portholes.

The second day passed tranquilly. The *Ben Franklin* moved at times, gently, because of the tide, yet maintained perfect stability. Half for fun and half on a wager, we made a house of cards that could be constructed as well as on terra firma.

Through the window the same fishes were visible: from time to time a ray passed by close to a porthole, also the little jacks and the plankton. For a harbor, the marine life was remarkably active. On resuming the watch on the second night I decided to dry out one of the hemispheres; equipped with a sponge, I removed more than a liter of water! Naturally enough, the moisture quickly condensed again on the wall, which was somewhat cool, but the condensation reduced the quantity of water in the atmosphere of the submarine. At least in principle, this is the best dehydrating mechanism, and in fact for several hours the "interior rain" stops. Since only a small part of the hull is accessible for this operation, the result cannot be complete: obviously only silica gel can solve this problem; unfortunately there was not enough aboard.

In the early hours of the morning the water suddenly acquired a magnificent clarity. Tens and probably hundreds of jacks surrounded us; the little ray, if it was the same one, was still there. A handsome sole glided along the bottom. Spot fish even come, apparently to clean the portholes by gobbling up the tiny shellfish that are trying to grow there. With Harold Dorr, we decided to keep the floodlights on for several hours: the fish and the plankton increased in density. We were in a veritable aquarium, but the reaction of the fish, or the absence of reaction, was difficult to interpret: for example, a brusque movement with a large flashlight or flashes of light coming from the interior of the mesoscaph did not seem to affect them in the least. Toward 4:28 a.m. a tug passed over the surface in the harbor and all at once the water lost its translucence.

By the third day certain habits and practices had firmly established themselves; nevertheless we were anxious to emerge, principally because of the humidity. As for the temperature, it was comfortable, or at least would have been comfortable if the air had been drier.

Toward noon two big barracudas, about a meter long, passed in front of the portholes. Then a bit later the little ray reappeared. The large number of portholes serve the purpose for which they were intended: one can follow the fish as they swim around the mesoscaph from porthole to porthole. At 3 p.m. I discovered that I was slightly out of breath, simply from having opened successively the various panels in the flooring.

This is a small routine task that requires no great effort. There was certainly more than 1 percent carbon dioxide in the air; shortness of breath is one of the first symptoms of the presence of excessive carbon dioxide. Our various detection devices still did not perform satisfactorily. There are three of these: each gives a different reading, and, moreover, these readings are not consistent from one moment to the next.

Ray Davis made his checks, taking constant and meticulous samplings.

In the afternoon I proposed that the surface send us some oranges that can be put in the airlock by a diver. It would do us good to have something natural to eat. Our secretary, Madame Simone Treyvaud, went to buy them and at 3:15 p.m. six fine Florida oranges came through the airlock in a plastic sack, arriving in perfect condition inside the mesoscaph.

Then came the third night. We were a bit fatigued, to be sure, but morale was high. We understood each other very well aboard, though I planned to recommend to Grumman that they establish a somewhat more flexible, less military discipline. I had the fixed intention that each one should profit widely by the dives that we are going to make at sea, and for this it is necessary that the atmosphere remain friendly and open. The discipline aboard a naval submarine is one thing, perhaps a necessary thing, but we are on a research submarine, and in my view this dive was somewhat hard to take.

During this last night, even at high tide with the water clear, I saw practically no fish, but at 12:57 a.m. Harold Dorr suddenly called me, in a voice muffled partly from excitement, partly out of respect for the sleep of others.

"Shark!"

I did not see it myself, but Harold described it to me. It was small, about 50 centimeters long, and was moving rapidly. Was it the shark's presence that dispersed the fish? A little later the fish peacefully reappeared, soles, rays, catfish, and various small alevins.

Cameron Walker, one of the men on watch on the surface, arrived in a boat exactly above the mesoscaph; he was trying to catch fish, and Dorr saw his hook and line outside the porthole. The fishing expedition was a disappointment: the hook attached itself to the mesoscaph—a magnificent catch, but too heavy, and Walker had to abandon it.

It is with such small games that oceanographers pass their time at night 11 meters deep on the bottom of a harbor in Florida. What did we see during these three days spent in the port? Barracudas, sharks, rays, soles, catfish, numerous and varied jacks, clouds of plankton, more than one often sees on a dive in a bathyscaph to a great depth.

Finally on the fourth day, or at the end of the third, depending on how you count it, in short at 9:50 a.m., Erwin Aebersold blew out the ballast tanks and brought the *Ben Franklin* to the surface. We did not open the doors yet, for we wanted the interior atmosphere to remain unchanged for the full seventy-two hours called for in the program.

Meanwhile the mesoscaph was towed to a wharf; some final tests were made and at 11:25 the equipression valve was opened. As the inside gauges had in fact indicated, we were still distributing a little more oxygen than we were consuming and the inside pressure had increased slightly. Opening the valve by equalizing the interior and exterior pressure produced, just as in a Wilson cloud chamber, the sudden appearance of fog, so thick for a few instants that visibility inside was practically zero. At 11:30 a.m. the hatch was opened, we emerged on the bridge and then onto the wharf where our families and Dr. Robert Jessup, Grumman's chief medical man, were waiting for us. A rapid medical check led to the conclusion that we were still alive. The dive had been happily completed and one more step toward the Gulf Stream had been taken.

22

NEW TESTS WITH THE *BEN FRANKLIN*

Much was still to be done before we could leave for the Gulf Stream. In particular, the *Ben Franklin* had to pass numerous "examinations." There had to be tests of towing, the first dive at sea, successive dives leading up to the maximum permissible depth, 610 meters, tests of the propulsion motors, of the endurance and capacity of the batteries; tests involving the dumping of the safety ballast at medium and extreme depth to see how the submarine behaved during a very rapid ascent. Successive dives would test the strength of the hull, and finally so-called official dives, with representatives of ABS who would give us the final certificate declaring that the mesoscaph had been constructed according to standards acceptable in America and that it was officially authorized to dive to a depth of 600 meters, and then with representatives of the U.S. Navy who had to assure themselves that the mesoscaph was safe enough so that, if need be, Washington could authorize its personnel to participate in the dives and that, moreover, the crew of the mesoscaph was competent.

Considering the large number of small research submarines and the desperate court their builders pay day and night to the Navy, bombarding the Pentagon and the various branches of the Navy with invitations to participate in free dives of one sort or another (in the hope of a fat contract to follow), the responsible authorities have had to establish a general policy. At the time of the *Trieste,* in 1957, simple confidence was enough, all the more so since there was no competition; the builder of the *Trieste,* Dr. Auguste Piccard, had supplied enough proofs so that no doubt was possible about the safety of his vessel. But much water has run under the bridge and now the law is of a Draconian severity permitting of no exceptions: no one connected with the Navy, even as a civilian, is allowed to dive in a civilian submarine that has not been "certified" by the Navy.

I have no intention of describing in detail each of these dives, though each one had its purpose, its charm, its individual interest. There were some forty of these before our departure for the Gulf Stream expedition, some of them only a few hours long, others a day or two, some in the harbor at Palm Beach, others off the coast or at sea between Florida and the Bahamas.

The first expedition at sea, January 27, 1969, involved a shallow dive. Since the continental shelf extends quite a distance off Palm Beach—that is to say, the bottom slopes down very gradually—we had to tow the *Ben Franklin* for two hours before we arrived above a suitable spot with a depth of 25 meters. The surface waves were moderate but had come from a long distance and could be felt even at 25 meters depth. The mesoscaph rocked on the sand and we stayed under water for only one hour that day. It was hardly more than a second christening.

On February 5 we were finally able to undertake more serious work. Two hours after leaving the harbor, having proceeded more rapidly than on the prior occasion, we found ourselves above a bottom close to 150 meters deep.

1:50 p.m.—the floodgates are opened.
1:58 p.m.—the dive begins.
1:59 p.m.—we are at 15 meters.
2:00 p.m.—we are at 30 meters; the *Ben Franklin* has never descended so far before.

At this depth in the open sea the water is a beautiful, deep blue; the sun is still present, directly, but its red rays have already been absorbed. There are no fish, but here and there specks of plankton are clearly visible. The sonar is activated. This is a machine that sweeps the area in front of the mesoscaph with an ultrasonic beam whose echo from any obstacle will warn us immediately of danger. Now we see numerous "targets" 100 meters distant. Not yet being acquainted with the instrument, none of us can say exactly what these objects are; but they are probably fish, perhaps the great game fishes—swordfish, marlins, barracudas—dear to Florida anglers. Though we see nothing through our portholes, the surface tells us that a shark is circling about our tender, the *Griffon,* that ancient trawler that The Real Eight Company has put at our disposal for these dives. One of the big thyristor inverters is showing signs of weakness. The captain prudently decides to return to the surface to fix it.

At 3:00 p.m. we begin to descend again.

At 3:02 we are at 60 meters. Don Kazimir requests permission from the surface to descend to that depth, and the *Ben Franklin* resumes its descents—40 meters, 50 meters, 60 meters, 100 meters. A little crab of some sort, a few centimeters wide, moves in front of our porthole. How vast the sea must look to him! Long chains of plankton drift in front of us. It is much darker now. We are near the bottom at 165 meters depth; it is beautifully calm. But we cannot touch bottom yet. First we have to rise to the surface. Too bad! Nevertheless we have been able to see that the doors and the great bolted joint of the hull are perfectly watertight.

At 4:00 p.m. we are once more floating on the surface. After an hour on the surface, we begin our third dive of the day; the descent lasts a quarter of an hour. This time the *Ben Franklin* is authorized to come to rest on the bottom.

The water is clear, but it is already quite dark. Of course, it is winter and this is late afternoon. The view is fine. At last I am again in beautiful, clear water in the real sea. Life is abundant here: lots of crabs—spider crabs, 10 to 15 centimeters across—and small cuttlefish as well. There is some current; we are on the edge of the Gulf Stream. A glance at the compass shows us that we are drifting toward the south. Is this a tidal current? Is it a local eddy? Is it a true countercurrent about which a few observers have spoken, among them those aboard the Reynolds submarine *Aluminaut*? It is hard to say; it is even hard to estimate its velocity, for we do not yet have a "current meter" aboard. Don Kazimir estimates it at 3 knots; Erwin Aebersold at 2 knots. An engineer we have aboard, Alan van Weele, a demon for precision, agrees with Don Kazimir at 3 knots; for my part I do not believe that the velocity is more than ¾ knot, or at most 1 knot. A half hour later, the current has changed direction. Now it is coming from the east. Our standard manometer indicates 160 meters. Taking account of the corrections for the compressibility of water and, even more important, for its salinity, that works out at about 152.4 meters. Al van Weele has installed an exact manometer based on the principle of the strain gauge—that is to say, measuring the mechanical tension and then the variation in the electrical resistance of a bar subjected to the pressure of the water. Thus it can estimate the depth to within a few centimeters. According to him, the center of the mesoscaph is at 151.8 meters.

At 8 p.m. I take the watch with Harold Dorr while the others get a few hours sleep. Nothing noteworthy except a whole school of cuttlefish 10 to 20 centimeters long, and once more a number of fine crabs strolling slowly over the sand, leaving the delicate imprint of their claws. Probably they are blue crabs, which live by preference near the mouths of rivers and descend to deeper waters during the winter. The extremities of their hindmost pair of claws are flattened in such a way as to facilitate free swimming. They seem to be disturbed if not frightened by the searchlights. When these are turned on, the crabs are very numerous in our field of vision, but little by little they withdraw and disappear; if all the light is turned off, they gradually return; when the searchlights are turned on again they are to be found all around the mesoscaph. They are not aware that they live from the sun, and this artificial light bothers them. It is the same with the little cuttlefish with their rapid, jerky motions. The squids continue on the hunt for bizarre little fishes; on several occasions I was able to observe a scene that repeats itself identically each time. A slender, almost transparent little fish is swimming along peacefully and suddenly it lets itself "fall," spinning, and buries itself in the sand. Then it re-emerges and at that moment is eaten by a cuttlefish that is waiting for it. Why? Each dive poses more problems than it answers.

The acoustic telephone is functioning well. A voice from the surface reaches us loud and clear. At 4 o'clock in the morning the current is coming from the north. I do not think that we are yet in the Gulf Stream; the perceptible current is a tidal one. It would be necessary to stay here several days to observe it properly.

At 4:30 a.m. it is my turn to take a nap. At 4:55 I am awakened. Don Kazimir has decided to return to the surface earlier than planned since one of the inverters, if not afire, has begun to smoke. First he woke up Erwin Aebersold to tell him, "There's a fire under your bed."

"Ah?" says Aebersold, taking up his bedclothes and moving to another berth where he continues his nap.

Since there is no hurry, we telephone the surface and ask permission to come up. It is, as a matter of fact, necessary to make sure the surface is free, that there is no danger of collision with some other vessel.

The *Griffon,* which constantly has an eye on us thanks to its acoustic detectors and the "ping" that we send out every two seconds, replies. It is the voice of Al Kuhn, the engineer of the watch, that we hear.

"Okay. The surface is clear. You can come up."

We begin by emptying the two negative tanks,

which reduces our weight by 750 kilograms. Then, according to the program arranged in advance for this day, we dump all the safety ballast to see how the *Ben Franklin* behaves during a rapid ascent. The ascent is indeed extremely rapid, more than 2 meters a second, but the most complete calm continues to surround us: not a jar, not a vibration. At 5:21 a.m. we break the surface; at 5:24 we are on the bridge. It is a fine night, in the distance the lights of the *Griffon* are clearly visible. In half an hour day will break; at 10 o'clock we shall be back in Palm Beach.

A week later we are off again, this time to carry out a different set of tests, at varying depths, relating to the solidity of the hull. I have already had occasion to mention this system, but it is worth a few more words.

In an electric circuit the resistance is proportional to the length (l) of the conductor and to a coefficient that varies according to the nature of this conductor; at the same time, it is proportional to the inverse of the cross section (s) of the conductor. For a given metal the known resistance therefore permits one to determine the relationship l/s. Now if one fastens a small resistor of known characteristics on a piece of steel, for example, this resistor will change its shape as the base is deformed under it with varying pressure; the electrical resistance will change too. Reading off this resistance from a Wheatstone bridge and taking into account certain corrections (especially for temperature) therefore gives the deformation, and this deformation or elongation makes it possible to calculate the stress, if one knows the elasticity coefficient, constant in the case of our hull. This method, becoming more and more widely used, is a great help to the engineer. Thus we had placed more than 400 of these strain gauges, some in the interior, some on the exterior, of the mesoscaph, in all the "critical" areas—that is to say, all those where an excessive deformation might be expected to produce serious results. This system was installed aboard by two Grumman engineers, Victor Hanna and Al van Weele.

February 13: the crew is aboard again. Don Kazimir, Harold Dorr, Erwin Aebersold, Al van Weele, an electrician named Ray Gregory, and myself.

The plan of the dive was to make a first descent to 100 meters and there read the strain gauges, ascend to the surface and reset the instruments at zero, make a second dive to 150 meters, and then one to 250 meters. To accomplish this we were to use our own propulsion system and navigate under water, heading eastward and following the slope of the continental shelf which brought us nearer and nearer to the great depths. Thus we had to cover a good number of kilometers under water, dependent on our own devices but always, of course, under the acoustic eye of the *Griffon.* This triple plunge gave us twenty-four hours of great beauty under water.

The grandeur of the sea, the richness of the ocean bottom, the perfection of the piloting progressively attained by our two pilots, Erwin Aebersold and Harold Dorr, and always the extraordinary calm, the submarine comfort which aroused the enthusiasm of each new passenger aboard the *Ben Franklin,* all these permitted us to return later to the surface with a new harvest of technical information and wonderful new memories.

At 1:55 p.m. and a depth of 60 meters we pass through a cloud of plankton, formed essentially of tiny jellyfish, most of them with long filaments. At 80 meters a small alert: water is running into the bilge! Don Kazimir insists that we ascend to 50 meters to see what is going on. It turns out to be merely the surplus of one of the hot-water reservoirs. All this water could run out and it would not change our equilibrium. The dive begins again; we reach the bottom at 115 meters. The bottom is scored with little ripple marks similar to, though less regular than, those one sees on beaches. Many small craters. Looking at them, one might easily believe we were on the moon. A little later, we ascend somewhat and set off again, all motors engaged, toward the east. This voyage is very beautiful. There is a great quantity of plankton, some handsome small jellyfishes, apparently of the genus *Gonionemus,* a few centimeters in diameter, shaped like a small, transparent bell from which hang dozens of tiny tentacles; these seem made of plastic with infinite delicacy, but zoologists tell us that they are 99 percent water. There are salpas, as well, a class of tunicates or urocords,—small, almost transparent. Most are individually separated, but some specimens are in chains, one is about a meter long. In the Mediterranean aboard the *Trieste* I once saw a chain of salpas at least 15 meters long and they are sometimes even longer. Suddenly we see coming toward us (actually we are approaching it) a gigantic mass of a kind of protoplasmic jelly, close to 0.5 cubic meter in volume, in the middle of which a very small jellyfish seems to be living. This enormous mass, which obviously does not weigh more than the water in which it floats—about half a ton—or nothing at all in relation to the water, is smashed against the bow of the mesoscaph without a sound, without a jolt, without a sigh, without a regret. The mass is nothing precise, a sort of secretion, no doubt, made almost entirely of water and held together by a few infinitely thin membranes; it is nothing but the expression of the richness of the sea: here there is only emptiness and life—and the distinction between the two often does not even exist.

At 3:00 p.m. we decide to descend again and reach bottom at 150 meters depth. A sea anemone, 15 centimeters long, some horseshoe crabs of various sizes: usually it is the males that are the smaller. Crabs decidedly abound in this region; a magnificent specimen passes rapidly in front of us. I will not say it is beautiful, with the parasitic shells attached to its carapace and its ten exaggeratedly long legs, but this spider crab nevertheless exhibits almost infinite grace in walking on the tips of its feet, and before disappearing it provides us with an entrancing ballet of the depths.

Night falls. A number of cuttlefish appear, along with small fish that, just as the other day, from time to time let themselves "fall" at full speed into the sand, which seems to swallow them up. When the searchlights are off one can see a few tiny points of light, but on the whole there is little bioluminescence.

At early dawn we set off again toward the east to reach a greater depth and to carry out new measurements. This time we rest on the bottom at 252 meters. The water is cold, 8°C., very clear, and plankton is scarce. By contrast we find a great number of minute crustaceans, apparently a kind of prawn. What are they doing here? Ah!—a long, dark furrow becomes visible on the horizon. What can it be? There are more and more prawns. The sonar does not indicate anything dangerous. The current is carrying us toward the furrow. Now that the perspective is better, we commence to see more clearly as we come nearer; it is a cable, one of those telephone cables that stretch from Florida to the Bahamas, as the chart shows us. All along the cable an enormous concentration of these tiny crustaceans. What are they up to? Are they aware of the vibrations passing along the cable? Are they listening to the conversations that speed by? Or, more probably, are they feasting on the polyethylene insulation?

11:30 p.m. The readings of the strain gauges have been completed. The hull at this depth, at least, is behaving perfectly, exactly according to the calculations made in Lausanne, Zürich, and Bethpage.

"Hello, surface! *Ben Franklin* here. Hello, surface! *Ben Franklin* here. Hello, surface . . ."

The surface does not reply. Our orders are explicit. If contact is lost we must ascend. We blow the water out of the VBT, to rise very gradually.

12:30 a.m.—still no reply from the surface.

12:45 a.m.—we are at 80 meters depth. The water is growing warmer. It is now close to 20°C. At the surface, radio contact is established. The *Griffon* has had trouble following us. (The tracking system still needs improvement.) It went a short distance away from the spot directly above us so that our calls, like theirs, were diverted by the thermocline, that zone where the water temperature changes abruptly and where acoustic waves are refracted or reflected like light waves in the hot air above a tarred road or the burning sands of the desert.

These two days have been very full. We will return to port to prepare for another expedition.

23

THE SURFACE DOES NOT REPLY

The next dive took place five days later. To finish our program of testing the hull, we must now descend once more to 250 meters (to recalibrate our instruments), then to about 500, 550, and 600 meters.

At our departure on February 18 the sea is far from calm. We are all eager to begin the dive. Aboard the *Griffon* especially, it promises to be far from comfortable during the next twenty-four to thirty-six hours; the *Griffon,* it will be recalled, is a small trawler, one of those boats that are unsinkable when piloted by trained seamen, even in the teeth of a tempest, but which bounce like cockleshells when the sea is not glassy smooth. Besides the crew of the Real Eight vessel there is the crew from Grumman, especially Vic Hanna, the technician who presides topside over the testing of the hull and who at each stage has to give his agreement and authorize us to go deeper.

At 260 meters it is already almost night, although the time is only four in the afternoon; a first pause for the strain gauges; for this we remain in equilibrium under water. The instruments are calibrated. Toward 7:30 p.m. we come to rest at 359 meters. A new reading of all the strain gauges. While Al van Weele begins his calculations of the deformation, I cast a glance through the portholes at a handsome echinoderm which I cannot precisely identify but whose phylum is recognizable by its radial symmetry; a few small crustaceans, but not much else. The sea temperature is 8° C. Inside the mesoscaph it is 15° C., which is at least bearable. Thanks to the silica gel, of which this time there is an abundance aboard, the relative humidity is only 70 percent, and the cold is easier to stand. We take a sample of sea water; its density is 1.027 at 13.3° C.

The plan had been to spend the night on the bottom with a normal routine watch and to resume work next day. On a proposal by Don Kazimir, however, we unanimously decide to work all night and finally next day to attain the depth of 600 meters which will be in some measure a demonstration of the soundness of the mesoscaph. Therefore we ask permission first to descend to about 500 meters. On receiving the okay from the surface we rise a short distance to resume our course toward the east. Van Weele works at his calculations. For the moment all goes well.

11:00 p.m.: we begin to descend again. A magnificent shrimp of 8 to 10 centimeters hangs motionless in the water, "suspended" by its long antennas as though from a gibbet. At 11:15 we reach bottom again at 495 meters. We take a rapid reading of the strain gauges and an hour later we are able to rise to 200 meters and resume our voyage in search of a depth of 550 meters. Everyone is at work, Van Weele on the hull, Erwin Aebersold and I taking turns at piloting, our eyes on the manometer and the echo sounder; when the total of the readings of these two instruments indicate 550 meters we will be able to go down again—if the surface permits.

At 2:00 a.m. we have found the place. At 2:15 we reach bottom: 538 meters. A reading of the strain gauges, a visual check of all the passages of cables, piping, the airlock, the great central joint. All is going well; even before Van Weele has finished his calculations we are ascending again in preparation for our search of the 600-meter mark.

"Hello, surface, hello, surface!"

The surface does not reply. The telephone is inexorably silent. We turn the mesoscaph in every direction to send the cone of our acoustic telephone toward all the azimuths. There is nothing to be done; the surface does not reply. The ocean is stormy; it is pitch dark; it would be the purest folly to ascend, whatever the theoretical instructions established in advance may be. And since we cannot return to the surface, it is just as well to continue our program and check the hull at 600 meters, the depth limit for which it was constructed. Everyone is in agreement; we are going to descend to 610 meters.

While we make our way at 2–3 knots toward the east where the greater depths lie, Van Weele feverishly completes his calculations. There are no suspicious points. When we arrive at the proper place, he will give us his consent; and after a final attempt at the telephone, which continues to be silent, we descend.

At 4:58 a.m. on February 19, 1969, the mesoscaph comes to rest at 612 meters. The hull is holding, obviously; not with absolute rigidity, of course, but all the strain gauges show readings in conformity with the theoretical studies and the models. Giovanola will be happy.

Forty minutes later we blow the water out of the variable ballast tank (VBT); obedient as a choir boy, the *Ben Franklin* ascends. We pause at 400 meters to try the telephone. In vain. It is very cold now (12° C.). We are in a hurry to reach the surface.

At 6:30 a.m. we are at 160 meters. The underwater day is beginning to dawn. There are some fine sonar "targets." For a quarter of an hour one of them seems to be following us at 200–300 meters distance. Is it a shark? A porpoise? Another submarine observing us?

At 7:15, after listening carefully on the telephone and turning our sonar in all directions, we break the surface. The sea is stormy, worse even than yesterday. Don Kazimir, Ray Gregory, and I climb onto the bridge. The view is splendid, impressive. There is a swell at least 4 to 5 meters high and, on top of that, heavy waves that break against our ballast tanks. The *Griffon* ? Not in sight. We are alone on the surface with nothing between us and the horizon; even the radio does not reply. . . .

For its part the *Griffon* has been looking for us. During the night it had lost its transducer because of the storm. Very soon afterward, the surface drift carried it beyond the range of our telephone. We can, of course, call the Coast Guard, who are always apprised of our dives. One of their helicopters could easily locate us and give our position to the *Griffon.* But there was no urgency and we continued to hope that the *Griffon* would hear us.

That is what finally happened. Once radio contact was re-established, we sent up rockets to indicate our position. But these rockets did not rise high enough. Under good conditions, at 30 meters they should have been visible to a distance of about 20 kilometers. Finally, with a radio goniometer, the *Griffon* found our direction and at last appeared on the horizon. At 10 a.m. the hauling cable was in place and the long tow began. During the first hours it was anything but comfortable, for on the surface the submarine rolled and pitched, if not as much as a conventional vessel of the same dimensions, at least more than one would like. Finally the sea grew calm; everyone could rest quietly in his berth. At nightfall we were still far from the shore; familiar sharp whistles and small barks could be heard in the sea. I got up. Through the portholes I saw four porpoises gamboling around us as if they wanted to assure themselves that all was well aboard. I picked up the microphone and tried to reply. There were a few more barks, and then they left. The sea was calm; the mesoscaph rocked gently, docilely allowing itself to be drawn by the *Griffon* at reduced speed toward the harbor of Palm Beach. Silence reigned aboard. Everyone was now asleep or on the verge of it.

PART THREE

The Longest Night

24

DEPARTURE FROM PALM BEACH

When the date of departure had been fixed—July 14, 1969—there was still an impressive number of unknowns to be faced.

This dive would be the first time the full crew would be aboard together, the first time the mesoscaph would dive in the open ocean for more than twenty-four hours. The entire crew this time would be subjected to a test of long duration. For the first time, the scientific equipment from the Navy, from NASA, in part from Grumman, would be tested aboard. We knew that the battery, which had been studied with considerable care, still presented problems, and there was no guarantee that it would behave properly for so long a period. In short, everything was perhaps fine on paper, but many tests might still have been made to advantage. We were far from the 99.9999 percent safety attributed to NASA in its space flights. To be realistic, it would have been a good idea to plan one more dive in the ocean of several days' duration, actually drifting in the Gulf Stream, and to use this occasion to perfect the navigation and tracking systems. However, the speed of the drift being at least as great as the speed of the tow, and, if the weather were bad, even greater, a three-day dive would automatically mean three to six days of being towed back home, easily a total additional delay of twelve to fifteen days, considering possible changes, recharging the batteries and the compressed air cylinders, and general reprovisioning for a thirty-day dive. It was already mid-July, well into the hurricane season. There was no precise final date for our departure, but each day that passed increased the risk of being interrupted in midcourse by one of those famous tropical depressions.

And so I proposed—at first to everyone's horror—to leave for three days and to decide at the end of that time whether to continue the dive or stop it and start again officially two weeks later. The idea was not a pleasing one, for it prevented an official announcement of our departure, and this greatly complicated the task of public relations. Moreover, it was asked, why prepare the mesoscaph for a month if the dive was to last only three days? On my side, I pointed out that the mesoscaph had never been provisioned for an expedition of a month and that there remained, for this if for no other reason, a good many things to try out, and a general rehearsal would teach us much that would be useful for everyone. If all went well, it would no longer be a general rehearsal but a successful operation all along the line. Finally, we were not, after all, setting out for the press or the public, and however pleasant it might be to be accompanied for some hours on the surface by friends in yachts, as was planned by many, the principle of the three-day rehearsal seemed to me more valuable still. The idea gained ground little by little, was accepted by the Grumman management, and finally was laid down as a new directive for all concerned. So it was that at 10:45 on the morning of July 14 we left as though for a routine dive, accompanied to the harbor only by a few guests and by our families.

The weather was fair, the sea calm, and towing presented no problem. There were only two men aboard to stand the first watch so as not to use up too much oxygen and, more important, not to produce too much carbonic acid gas and too much humidity. As a matter of fact, the whole air-conditioning system—that is, the maintenance of a satisfactory ratio in the air of oxygen, carbon dioxide, and humidity—had been calculated for a period of six weeks. Nevertheless we were aware that the calculation in certain respects was a bit theoretical or, more correctly, a trifle arbitrary, and we needed to take advantage of all possible margins. If we knew, according to the tables established and frequently recalculated, especially by NASA, how much oxygen a man required per minute, we could not know what that average would be for us, for it would depend basically on our physical exertion. For a man at rest, seated at his desk, let us say, 0.2 liter per minute is the minimum; 0.5 liter per minute is the average for moderate but semicontinuous effort. Consumption, therefore, may easily vary by a factor of 2 or even more. We knew how much humidity we were going to produce and theoretically how much silica gel would be required to absorb it. But here too there were several unknowns: would the little bags of silica gel become completely saturated? Or would their centers remain unaffected by the humidity in the atmosphere? If so, in what proportions? Above all, how much humidity would be produced by accessories such as the kitchen and the shower, for example? We had a good margin in that six-week calculation, yet we had to be prudent from the start. Moreover, these long tows were no treat for those aboard the *Ben Franklin.* Even when the sea is friendly the mesoscaph is ceaselessly tossed about; it rolls and pitches; the interior temperature climbs rapidly under the sun's rays which in Florida strike the bridge almost perpendicularly. All good reasons for confining the watch on the *Ben Franklin* that day to two men.

The tow lasted seven and a half hours. When we arrived at the position for the dive, we still had two hours of daylight to make final preparations. Since we could not know in advance what the sea would be like, some of the oceanographic instruments had not been placed aboard, and some, the current meter, for example, had been packed in such a way as to protect them against possible heavy waves. So the crew from Navoceano (code name for the Navy Oceanographic Office) went aboard first and worked strenuously to get everything in place as soon as possible. This was important; by agreement with the Navy it was necessary that everything should function impeccably at the beginning of the dive.

During this time the *Ben Franklin,* still attached to the tug, was drifting slowly toward the north since, obviously, we were already in the Gulf Stream.

At 8:25 p.m.—it was already practically night—the great nylon towing rope was cast off. Five minutes later I boarded the mesoscaph and at 8:34 on July 14 we closed the door of our prison: it was our own way of celebrating the fall of the Bastille.

25

TOUR OF INSPECTION

It may be useful at the moment when we begin this dive, to survey our habitation and to offer a few more details about our shipmates.

The interior of the mesoscaph, although not sumptuous, was nevertheless spacious and relatively comfortable. It lacked one aspect that in our first designs at Lausanne I had planned to give it: inner walls of mahogany and a general arrangement similar to the interior of a nice yacht. I had thought that a good warm wood with perhaps even some tapestries—let us say the interior of a gilded cage—would ease a long sojourn aboard. Though agreeing in principle, Grumman would not accept the idea of wood. When I proposed mahogany bulkheads, the notion was greeted with raised brows.

"But mahogany—that's wood!"

"Yes, of course."

"It's out of the question. It would catch fire too easily."

It did no good to object that there was no more reason for fire aboard the mesoscaph than on any yacht, and that wood could be treated to make it noninflammable. I also proposed using ultrathin sheets of mahogany cemented to metal, but this concession was not enough. It was suggested that a plastic imitation wood be used. In my turn I did not take to this compromise. No plastic imitation produces the effect of mahogany in European eyes, any more than stucco could replace marble under the chisel of the Greeks. If wood was excluded, we had to give up the appearance of a yacht and fit the inside of the mesoscaph with aluminum panels painted white. Because of this the interior of our submarine has been compared to that of a hospital: it had the same appearance of neatness, cleanliness, and, theoretically at least, asepsis. A long corridor runs from bow to stern, 80 centimeters in width, just enough for two people to pass. At the bow the wardroom, hemispherical in shape, forms the front end of the cylindrical hull. Along the walls of the hemisphere are seats whose backs constitute a number of lockers packed tight with bags of silica gel and sacks of dehydrated food. On one of the walls hung the chart of the region we would traverse during the first half of our voyage. In the middle of this "saloon" is a round table with a few chairs. This little room will serve as meeting place, office, dining room, living room, and workroom.

The long corridor is lined with various instruments, as well as by the berths, the shower, and the head. First of all, to starboard, there is a large locker jammed with electrical and electronic equipment: the recorder of ampere-hours and other devices for checking the batteries; the surface radio; the sonars; a television station connected to an underwater camera and a floodlight; the various electrical controls. Then comes the galley which consists of a small, stainless steel sink with flat surfaces of formica on either side and overhead four thermos tanks containing hot water for the duration of the mission. This water has been heated in advance with the harbor electricity, in order to save our batteries. Beyond the galley is one of the big AEG inverters, those devices that transform the direct current of the batteries into alternating current, primarily for our propulsion motors. This machine has above it a ventilator that distributes throughout the vessel the heat produced by the inverters: the heat is welcome during dives when the water is often so cold, but it actually represents the 10, 20, or 30 percent loss of output against which we fought tooth and nail with AEG during the designing of the electric circuits.

Next comes the shower. It is a small closed compartment the size of a standard shower stall, well insulated and watertight. As a matter of fact, it is smaller than originally planned; one of its walls is double, forming a separate compartment where some of the emergency batteries have been stored.

Beyond the shower, a double level; below, two portholes (one equipped with a device for taking samples of plankton which I shall describe later) and a locker housing Draeger apparatus which theoretically would allow us to escape from the mesoscaph should it be disabled on the bottom beyond the limit of possible survival aboard but at a depth of not more than 300 meters. In fact, the depth of 300 meters has never been attained at sea except by Swiss diver Hannes Keller, and I certainly hope not to have to leave the *Ben Franklin* by means of this rescue device. On the lower level are also some small electrical inverters for part of the equipment aboard. On the upper level is my berth, long enough but rather narrow; however, it has one special attraction: I have had installed a porthole 30 centimeters from my pillow. Stretched out, I can still see the ocean, better than from a hotel on Miami Beach. I will spend many hours lying on that bunk, staring into the infinitude of the seas, ready to seize upon the tiniest speck of plankton, the smallest fish, the least trace of life.

The compartment beyond my berth is reserved for the oceanographers: an upper berth is assigned to Frank Busby and beneath it is the oceanographic equipment. This berth is at the end of the corridor. Beyond it is the stern hemisphere, provided, like the forward one, with six portholes and an access hatch for reaching the bridge when we are on the surface. Returning along the starboard side of the corridor, there is, first, Ken Haigh's berth, below which is arranged an important section of oceanographic instruments which are immovable and record the nature of the dive; here too, as on the opposite side, are the Navy instruments which will be described later. Moving forward, there are two more berths, one above the other, those of Chet May and Don Kazimir; then beyond the head is another compartment containing the second AEG inverter; beyond the inverter a large compartment containing the principal electrical relays and commutators, and above them the sixth berth, the pilot's. The chief pilot of the mesoscaph, Erwin Aebersold, had often joked about his berth's being on top of a mass of relays, resistors, condensers, switches—in short, innumerable electrical gadgets. At least, he said, I won't be cold. He did not know how truly he spoke. One day, as mentioned before, he had to change his berth when a contact began to smolder and a small black plume of smoke emerged from under his bed.

Just beyond the pilot's berth is the pilot's station itself. All the instruments that can be brought together are sensibly arranged like those in an airplane, on two principal panels facing the pilot. At the top are the various electrical devices, the safety system, including the Longines timers, the controls for the explosive bolts that jettison certain exterior accessories in case of emergency; the acoustic sounder (a Norwegian device called SIMRAD which, at Aebersold's suggestion, was rechristened Sinbad), the lighted detectors showing the quantity of water gradually entering the oil reservoirs of the batteries in proportion to our depth, and other detectors to show the presence of water—a source of danger—in other exterior instruments. In the center of the panel are the Rolex chronometers and chronographs showing the official ship's time. Once more Rolex has offered its valuable collaboration and has designed for our expedition highly precise quartz timepieces: these will be of the greatest importance during the month in which no exact time signal from outside can be received.

Arranged on the lower panel are the electrical controls, properly so called: the levers and the various small switches controlling and regulating the speed, direction, and position of the four propulsion motors. On the pilot's left are the principal hydraulic and pneumatic controls. These actuate the compressed-air system which serves to blow the water out of the principal ballast tanks when the mesoscaph reaches the surface, or out of the VBTs when it is necessary for some reason to reduce the weight of the mesoscaph under water. In the midst of this is a series of manometers, among them the two principal depth indicators supplied by Haenni, the firm that had equipped the first bathyscaph, the *Trieste,* and the first mesoscaph. One apparatus, for small depths, can indicate a variation of a few decimeters; this device is automatically switched off at its maximum depth of 80 meters. Valid up to 1,000 meters is another manometer, graduated in 10-meter intervals but on which a variation of a single meter can be estimated.

Finally, behind the pilot is a recording manometer of high pressure on which one can follow the slightest variation in depth and in particular the rate of ascent or descent, which is indicated by the angle of a line traced in ink.

In front of the pilot's post is the forward hemisphere already described. This concludes the tour of the premises, except for a word about the floor and ceiling.

The floor, at 40 centimeters above the lowest level of the bilge, is covered with marbled beige linoleum given us by the Deutsches Linoleum Werk A. G. It is divided into removable panels, permitting easy access to the bilge where it is necessary from time to time to inspect cock sluices, a part of the tubing of the internal water system, the waste tanks, and the principal fuse panels enclosed in watertight compartments. Above, there is no ceiling as such; and in plain view are several pipes colored according to their functions: yellow for pressure oil, gray for air, blue for water, and so on; a porthole for vertical observation toward the surface:

the small airlock, already described, for releasing possible "messengers"; and a series of hull penetrations for electric cables and pipes connecting the interior with the exterior of the mesoscaph.

Most visitors, especially those familiar with the conventional submarine, are surprised by the free space and roomy arrangement of the *Ben Franklin.* The platforms at each end and the central corridor give the impression of spaciousness, but above all the twenty-eight portholes (the twenty-ninth having been given over to the device for taking samples of plankton) allow one to feel as if he is in the sea rather than in a steel prison. Psychologists know that windows in an airplane, an apartment, a boat, and here in a submarine, permitting a person to look outward, help to reduce the feeling of claustrophobia which blank walls can cause. People who are disturbed at half a minute in an elevator have no fear of spending half a day in an airplane.

Unquestionably, throughout the whole time of the dive that awaits us, these windows will be a precious link with the exterior.

26

SIX MEN IN A BOAT

Before pushing the buttons that will open the flood valves and begin the *Ben Franklin*'s descent, I would like to present the crew of the mesoscaph—this crew that will have to live together for a month, sharing all their impressions, emotions, joys, and perhaps fears, throughout all the 720 hours of the Gulf Stream expedition.

First the captain: a veteran submariner in both conventional and nuclear vessels, employed by Grumman two years before, Don Kazimir, although still young (at thirty-five the youngest member of the crew), has had years of experience at sea aboard the most modern and complex vessels and is thoroughly familiar with submarine technique. The mesoscaph, however, is the first submarine in his experience to have portholes. He knows the sea through the media of listening devices, hydrophones, sonars, and various ultramodern sensors, but in the *Ben Franklin* he sees the actual underwater world with his own eyes. He represents Grumman's authority aboard. He is responsible for the mesoscaph and he knows it very well, for he came to Monthey during the winter of 1967–68 to witness part of its construction and followed its reassembly step by step in Palm Beach. He has memorized the electric currents we worked out in Lausanne, Hamburg, and Monthey, and the hydraulic and pneumatic systems; he has learned the function of each piece of apparatus, all the ways of using the *Ben Franklin.* Also he has a quality that makes him greatly valued by his superiors: a soldier at heart, he never disputes an order. He is obedient body and soul to the "surface," the authority responsible for all operations. How much water has run under the bridge since the first dives of the *Trieste*! Now individualism has disappeared, its function replaced by groups and committees, with all the advantages but also the disadvantages that group action entails.

Erwin Aebersold, the pilot of the mesoscaph, is Don Kazimir's adjutant and also my principal assistant. Known as Erwin in America, Monsieur Aebersold in Lausanne, he is a pilot by nature, an airplane pilot trained in blind flying, past master in the use of flight simulators. He has worked with me since 1962 when I began the designs of the PX-8, the future mesoscaph *Auguste Piccard.* A first-rate builder, he designed in particular the pilot's station on the first mesoscaph, and he has done the same, among many other things, for the *Ben Franklin.* He has been on all the dives of the mesoscaph in America. In every situation, the vessel responds to his hand and eye.

Then we have two oceanographers. Frank Busby, young, debonair, energetic, graduated in oceanography from Texas A. and M.; a civilian employee of the Navy Oceanographic Office in Washington, he knows about all the research submarines in the world and shares in the secrets of the U.S. Navy concerning all new constructions of undersea research craft. Frank has two missions aboard: to explore the Gulf Stream and to explore the *Ben Franklin*—that is, to evaluate its possibilities and to note what may be useful for future civilian or Navy submarines.

Ken Haigh is our second oceanographer. A member of the British Royal Navy on detached service for two years with the U.S. Navy, a specialist in acoustics, he is our universal and active "listener." In his person the British empire has come aboard the mesoscaph: calm, level-headed, knowledgeable, dependable, humorous, discreet, modest, tenacious, impassive—all these British qualities go into making a wonderful coworker and comrade. Only one thing aboard displeases him: the instant tea! I agree with him.

Chet May, a NASA engineer and our fifth member, is an observer. In 1972 or 1973 NASA plans to put into earth orbit a large, permanent laboratory which small teams of scientists will take turns in manning for several weeks at a time. Saturn rockets and special capsules will provide an almost regular shuttle service. How will the scientists live on board? What space will they need—not just for their scientific and technical equipment but for themselves, their books, their distractions? What will they eat and how much water will they need for drinking, for washing, for their various experiments? How many hours out of twenty-four will they want to devote to work, to sleep, to recreation, to exchanging ideas? How much time will they spend looking out of the portholes? How many portholes should be planned? What about the "biological" life—what will the bacteria, microbes, and viruses be up to? How can dishes be cleaned with very little water? How manage disinfection if an epidemic should break out? Here is a whole new field of exploration wherein our voyage offers a marvelous opportunity, making Chet May our "life" engineer. During the whole dive he will observe us with the eye of an expert, however discreetly. Three automatic cameras aboard will photograph us every two minutes: that adds up to 64,800 photos. Under our berths are hidden counters that will record to within a few minutes how much time we will have spent there. On the floor other counters will register our steps. Regularly Chet will examine our "biological state," taking samples from our skin from the washbasins, the toilets, the floor, the ceiling, the portholes, and getting them to grow in cultures for later examination in the NASA and Grumman laboratories. Each day he will have us take part in an electronic game that records the rapidity of our reflexes to discover whether they are being modified for good or ill by the long, very long, isolation. This is called the Space Skills Test. Naturally enough we nicknamed it the Space Killer.

Including me, with the somewhat pompous title of "leader" of the expedition, that makes six—six men in a boat.

27

THE BEGINNING OF THE DIVE

At the moment of entering the mesoscaph, I find the interior almost as bright as day: all the electric lights are turned on and the illumination contrasts violently with the descending night outside. Rapidly Don Kazimir and Erwin Aebersold run through the final check list, each item of which has to be verified by at least two members of the crew.

8:35 p.m.: it is almost a minute since the hatch was closed. We distinctly hear "someone" rapping on the outside. Is it a diver? A rope tossed about by the waves? We shall never know.

Don Kazimir—generally called Kaz in the American fashion—is on the underwater telephone. This is a device that transmits the voice through water in the same way that radio does in air and space, but using acoustic waves instead of Hertzian, which do not travel through water except in certain special cases. With our telephone a carrier wave of 8.087 Kilohertz transmits frequencies of 8.3 to 10.7 Kilohertz, which permits us in addition to hear a whole series of sea sounds other than those directly intended for us. Kaz is calling the *Privateer,* the escort boat, which the Navy hired for us from Reynolds International, the aluminum company that built the research submarine *Aluminaut.*

"*Privateer, Privateer,* this is *Franklin.* Over."

(Gurgling.)

"*Privateer, Privateer,* this is *Franklin.* Over."

(More gurgling.)

"*Privateer, Privateer,*" Kaz goes on repeating for three minutes, and only the gurglings of the sea reply.

We look at one another, mildly anxious. We have four telephones available aboard; at the start all four should be functioning perfectly.

Don picks up the radio, establishes communication, and wants to know if we have been heard on the telephone. The reply is negative but reassuring: those on the surface, still seeing us in the light of their searchlights, have not even turned on the underwater telephone.

At 8:40 p.m. everything is in order. We can open the flood valves; the sea is fairly calm, but we will nevertheless be more comfortable under the surface. The mesoscaph rolls in the waves and the sensation is not exactly agreeable. Usually, to accelerate a dive, we turn on the diving motors in order to leave the surface faster; this time, to economize on our batteries, we do not do so. Also, our rate of dive is regular but somewhat slow: if the mesoscaph has been properly weighted in relation to the temperature and density of the water in that place, it will take fourteen minutes to disappear entirely. I am in the forward hemisphere; through the porthole in the door, my eyes are fixed on the conning tower which still stands out clear against the night sky. Aebersold is at the pilot's station, Kazimir beside him. Everyone has work to do.

8:54: the conning tower is completely submerged. The loading and the distribution of weight for departure were perfect; the dive has begun.

It had been planned that the *Ben Franklin* should first of all spend part of the night on the bottom, 500 to 600 meters down. This would allow us to catch our breath and also to make a final general inspection, testing all the technical and scientific equipment, and it would give Frank Busby an opportunity to study the sea bottom off Palm Beach. A number of other locations for setting down had been picked out, at each of which, circumstances permitting, we were to spend several hours on Busby's behalf.

The *Ben Franklin* had been loaded with enough iron shot to take it directly to the bottom. The plan was to dump this shot in three installments: a first during the descent to make up for the excess ballast intentionally loaded at the start to make sure that we would descend, a second lot during our stay on the bottom to make up for the cooling and the resulting additional density, and the final third in order to ascend to a depth of 200 meters, which was to be our average drifting depth. Once this was accomplished, the remainder would be simply emergency ballast.

Our speed of descent, 20 meters a minute to begin with, was good. We let the mesoscaph descend peacefully toward the bottom at a relatively constant speed which automatically integrated, like some superelectronic computer, a whole series of varied phenomena: the gradual cooling of the water outside (30° C. at the surface, 17.68° at 195 meters, 7.33° at 420 meters, 6.65° on the bottom) which tends to diminish the speed of descent; compression of the gas in the exterior batteries (which tends, mainly at the beginning, to accelerate the speed of descent); compressibility of the ocean water (which tends to slow up the descent); compressibility of the hull (which tends to accelerate it); change in the salinity (which can either accelerate or slow the descent, depending).

At 70 meters, three minutes after leaving the surface, we already see abundant plankton, brilliant in the beam of the searchlights, streaming upward outside the portholes like snow snatched skyward by a squall.

Toward 300 meters the speed is already reduced by half, a direct illustration of the basic principle which determines the stability of the mesoscaph. But this rate is still too fast for a soft landing on unknown bottom.

It is now 9:27 p.m.; we are at 450 meters. The fathometer indicates that the bottom is 60 meters below us: time to put on the brakes. We dump iron shot in four lots for ten seconds each, a total of 400 pounds, less than 200 kilograms. At 9:48 the guide rope—a cable, copied from the balloon, which hangs below the mesoscaph—touches bottom before we do and automatically brings us into equilibrium by freeing the mesoscaph of its weight. Dragging along the bottom, the guide rope stops our descent, as it is supposed to do, at 10 meters above the ocean floor. Since the guide rope is of course attached to the stern of the mesoscaph, it serves to orient us in the current if there is one; here the Gulf Stream apparently does not make itself strongly felt, nevertheless in a few moments we are gently proceeding straight forward at an estimated speed of 1/10 knot.

The water is exceptionally clear, as it often is at the bottom. There is hardly any plankton about at the moment; the only things in sight are a handsome shrimp floating at the height of our central portholes and on the bottom a small anemone and occasionally one or two tiny fish, probably Myctophidae (lantern fishes).

The interior temperature is agreeable. It is now 26°C. as against 29.5° at our departure, which was too high. On the other hand, the humidity has greatly increased: 79 percent as against 54 percent when we shut the door.

Frank and Ken check their instruments and prepare for the first readings. Chet arranges his material. Erwin is at the pilot's station. Kaz is doing the cooking; he is trying to find out how to prepare dehydrated foods so that they will look, if not appetizing, which would be too much to ask, at least passable. Finally he picks up the telephone:

"*Privateer, Privateer,* this is *Franklin.* Over."

The reply comes in a few seconds, clear, strong, and hollow, surrounded by scattered echoes:

"*Franklin,* this is *Privateer.* Over."

"*Privateer,* this is *Franklin.* Please tell me how many cups in a pint. Over."

We depend on the surface for everything, even for our cooking.

At midnight we are all still on our feet. The day has barely begun. The water is so limpid, so clear, that it is worthwhile to take photographs.

But it is beginning to get quite chilly; 20.5°C. Some of the humidity condenses on the walls. If we wish to, we can mop up this water with sponges to keep it from evaporating when the interior temperature rises again. This was part of the plan, too, in case the silica gel was not sufficient: to descend from time to time into cold water precisely in order to condense and collect the humidity and then re-ascend with the interior air warmer and drier.

Anyway, the *Ben Franklin* has been considerably cooled. This has increased its weight in relation to the water, and it rests for some hours on the bottom, the guide rope extended at full length behind it. This is the moment that Frank Busby has chosen to proceed with a first series of records of acoustic waves artificially produced on the surface. At his request ten explosions are set off at one-minute intervals; these are percussion caps electrically detonated by the *Privateer.* The waves come and strike our listening devices, the hydrophones placed on the outside of the mesoscaph. We hear them clearly enough without any instrument, and also various echoes, for they ricochet from the surface, from the bottom, rebound upward, descend again toward us, and thus leave a multiple record on Ken's magnetic tapes. Before declaring himself satisfied, Ken runs back the tapes and *watches* each explosion march past on a cathode tube. Everything has gone well, we can resume our course with the Gulf Stream. Before leaving, we glance at the "current meter." It indicates 0.2 knot. This, then, will be our approximate speed when once more we float in equilibrium above the guide rope.

July 15. Precisely at 1:00 a.m. I dump 50 kilograms of ballast, then in eight successive lots 100 kilograms more. We have just gently risen when we see passing by a magnificent crab some 30 centimeters wide.

One must be careful in estimating dimensions under water. Without a porthole correcting lens everything appears nearer than it actually is because of the refraction of light, in the proportion of four to three, and, depending on the searchlights used, the actual distance is hard to estimate. Often one finds himself wondering whether a fish observed is small and near at hand or large and distant.

Here, however, the bottom does not seem to be particularly rich. Nevertheless I have splendid memories of that first slow drift with the current. From time to time in the crystalline limpidity of the sea, one can see a shrimp, a few small tropical fish of the angelfish variety, a handsome ray swimming gracefully along the bottom deploying some 30 centimeters of marvelously supple fins.

The current meter reads zero, which shows that the friction of the guide rope on the bottom is negligible; we are proceeding at the speed of the current. A slender fish approaches, no doubt a rat-tail fish, common at this depth. Some beautiful large sea urchins move slowly along; they are probably 12 to 15 centimeters in diameter.

At 2:15 a.m. we observe that the specks of plankton surrounding us are moving a little faster than we are. The current meter, moreover, indicates a relative speed of 0.5 knot. We dump a little more iron shot in order to rise a bit on the guide rope and regain the speed of the current.

It is cold, 17.5°C. We are moving steadily, our bow searchlight turned on; this is one of the new searchlights with thalium whose greenish light is particularly penetrating. We are also preceded by the acoustic beam of the sonar called the CTFM (Continuous Transmitting Frequency Modulator); it sweeps the sea from side to side in front of us to a distance of 1,400 meters with the duty of informing us instantly of any possible obstacle.

At 2:28 a.m. another sea urchin; at 2:42 still another; at 2:45 a brand-new can of preserves, brilliant in the beam of our searchlights. The current is taking us toward the north-northeast at exactly 025° by our ship's compass. We are at 8 meters from the bottom, drifting steadily. The humidity has fallen to 54 percent. I decide to go to sleep for a few hours. The day has been good, fruitful, but long.

During the rest of the night on four occasions 5 kilograms of shot are dumped. Thanks to this, the depth and altitude of the mesoscaph have remained constant to within about 4 to 5 meters. The bottom itself is practically flat, and the drift continues without incident. But the interior temperature has dropped again: it is now close to 13°C.; the humidity is satisfactory, 62 percent. Since the sea is at 6.59°C., if we remain here a few hours longer the interior temperature will drop to around 7.5°C. Fortunately all the observations we wished to make during this first stage have been completed and we can begin our ascent toward 200 meters, which is to be our average depth during the drift.

28

LOG

Here are the notes from my log, reproduced just as I made them, with a few explanatory comments in brackets:

9:48 a.m.: 10 kg of shot—we ascend a little.
9:54 a.m.: 10 kg of shot—we ascend a little.
9:58 a.m.: 10 kg of shot—we ascend a little.
10:10 a.m.: 150 feet above the bottom. Chet May doing gymnastics [principally for warmth].
10:10 a.m.: [depth] 465 m. [That is to say, we have already ascended 45 meters].
10:18 a.m.: 461 m. Direction [of mesoscaph in the current] 135°. 6.98°C. [temperature of

the ocean]. 6 seconds—50 m. [that is, 6 seconds drop of ballast, or 30 kilograms, have made us rise 50 meters].

10:27 a.m.: 15 kg of shot.

10:29 a.m.: 460 m. 6.99°C.

10:31 a.m.: 450 m. 7.00°C.

. . .

10:43 a.m.: 409 m.

10:45 a.m.: 407 m. 7.23°C.

10:54 a.m.: 405 m. 3 seconds [blowing out air from one variable ballast tank has made us rise by] 45 m.

11:03 a.m.: 407 m. I let it [the mesocaph] oscillate a little.

11:12 a.m.: 410 m.

11:13 a.m.: 15 kg of shot.

11:25 a.m.: 393 m.

11:33 a.m.: 393 m.

11:38 a.m.: 393 m. 8.30°C. 35.05 percent salinity. 1494.5 m/sec [speed of sound in water]. *Privateer*'s propeller can be heard [on the telephone].
[One of the Navy recording devices periodically emits a bizarre sound like] someone snoring. (We will hear this strange sound for thirty days and quite often mistake it for a real snorer.)

. . .

11:45 a.m.: 62 percent (humidity) 12.8°C. [interior temperature]. 392 m.
[It is cold. Aebersold is taking a few flashlight photographs.] We warm our hands on the flashbulbs.

11:52 a.m.: 391 m.
12:05 p.m.: 395 m.

This ascent is carried out exactly as planned. We rise gently to allow the mesoscaph time to acquire the approximate temperature of the water. It would have been easy to rise much faster and to pause at 200 meters. But then the mesoscaph would have retained the temperature of the bottom water; warming gradually, it would then have expanded, its density would have diminished, and pretty soon it would have begun to rise again. At this point we would have had to take on water in the variable ballast tanks (VBTs), and that would have reduced by that much our margin for maneuvering in subsequent operations. We are anxious to start the real drift at about 200 meters with our two VBTs empty.

Besides, we are in no hurry. The sea is still beautiful although with all the exterior lights turned off there is nothing in particular to see; not a trace of bioluminescence for the moment, no plankton or phosphorescent fish. At 400 meters there is still some light, especially if one looks toward the surface through one of the "vertical" portholes; even looking toward the bottom you can still see a certain glow in the water. We know from observations made by the *Privateer* that we are in the center of the current, in its "core." We are told that even at 275 meters we will find the water at 15°C. and that at 175 meters it will be 18°C.

The surface regularly measures the water temperature with what is called an "expendable bathythermograph," a modern device that is replacing the old bathythermograph which has rendered such great service since its invention. It is called "expendable" because it is intended to be sacrificed, discarded. This is an electric thermometer in a small plastic capsule which also contains some hundreds of meters of electric wire, one end of which is connected to the thermometer and the other to recording apparatus aboard the vessel on the surface. The observer throws the thermometer into the water and the wire unrolls, transmitting to the boat the temperature of the water where the thermometer is. When all the wire has unrolled, the weight of the apparatus breaks it, and the thermometer sinks, lost forever. The whole curve of the temperature as a function of the depth has been registered during the descent of the apparatus.

It may seem wasteful to lose a thermometer with each measurement. As a matter of fact—and this is what has made the use of the new method spread so fast—if one takes into account the saving of time for the crew of the ship, the saving of fuel—for the ship no longer has to stop and start for each measurement—the saving of time for those who have to read off the results, this method is incomparably more economical and more efficient than the former one. The *Privateer* and the *Lynch* (the oceanographic tender that was soon to join us) have brought with them a thousand of these thermometers, which will prove of enormous value during our expedition.

29

LOST IN THE GULF STREAM

At 2:03 p.m. we are 328 meters under the surface. I suddenly observe an irregularly formed mass, between 20 and 30 centimeters long, which is descending and brushing one of our portholes. It looks like a carrot. It is probably a cuttlefish. Some minutes later we seem to be caught in a small eddy or whirlpool: in fact, Frank Busby observes from the compass and the direction of the particles, easily visible in the water, that we are drifting toward the south; this, however, lasts only half an hour. But another more serious problem arises: the acoustic telephone does not reply. Some vague sounds come from the surface, and clearly audible echoes repeat our calls. But no conversation is possible. At 3:14 p.m. we are at a depth of 300 meters. But the *Privateer* has lost us.

The situation is not dangerous, but it has its dramatic aspects. It has been decided with Grumman in advance that every half hour there should be a telephonic exchange. In general it would be a simple matter. The surface would call: *"Ben Franklin,* this is *Privateer.* Over." We would reply: "Roger. Out." In Navy slang this is called a "Gertrude" check.

As a matter of fact, the submarine telephone uses up precious current, especially, of course, during emission, and so although receivers are kept open on both sides, we, the *Ben Franklin,* talk only when strictly necessary.

An established rule exists that if no contact had been made in the course of an hour, that is, if the telephone does not reply and the tracking system, about which I will speak again, is not working, the *Ben Franklin has* to rise to the surface. This clause had been implicitly included in the agreements reached with the Navy and with NASA. Now we are in danger of having to apply it in less than an hour unless satisfactory contact is re-established.

In addition to our two usual telephones, of relatively short range but economical in their use of power, we have two others, much more powerful but devourers of ampere-hours as soon as they are turned on, and so the plan has been to use them only in cases of emergency. This is the time to try one of them. We had never used one during a dive; they had been installed aboard during the weekend, on the eve of departure.

"Privateer, Privateer, Ben Franklin here. Over."

The response comes, feeble but audible: *"Ben Franklin, Privateer* here. Prepare for an estimate of distance."

Immediately we hear, feebly, the voice from the

surface counting: "Six, five, four, three, two, one, zero."

At the precise instant of zero Don Kazimir cries: "Mark!"

The surface has started a chronometer at the word zero: the chronometer is stopped when they hear Don say, "Mark!" The chronometer registers the time required for sound to make the round trip from the *Privateer* to the *Ben Franklin* and back. Knowing the speed of sound (in practice 1,500 meters per second), the crew on the surface deduces the distance of the mesoscaph. Then it can draw a circle around its own position. The mesoscaph must be on some point of that circle.

Then, proceeding at full speed in the "probable" direction of the mesoscaph, it begins the operation over again: "Six, five, four, three, two, one, zero." "Mark!"

The new circle may cross the first one or, in the most favorable case, lie tangent to it. In the former situation there are only two possible points and a third measurement will decide between them. In the second situation the mesoscaph is located instantly. If the circles do not touch at all, the full check has to be started again.

The operation is simple in principle but it has to be done quickly, for the sea is not a drawing board, the Gulf Stream flows toward the north faster on the surface than at depth, and we do not even know with certainty in what direction we are traveling. Besides, the sea is not empty; there are other vessels on the surface to which attention must be paid. We found out later that the *Privateer* had lost us, not only because it lacked the training that it was to acquire little by little during the mission but also because it had been besieged for several minutes by radio calls coming from land and from one or more ships the presence of which had forced it to alter its course. But all's well that ends well; at 3:39 p.m. contact is fully re-established and tracking can begin again. Without the powerful emergency telephone and without the outstanding skill of the surface crew we would have had to ascend to the surface, re-establish contact by radio and visually, and then resume the dive; this would have been a most regrettable interruption, especially on the second day of the mission.

The word "tracking" calls for some explanation. It is clear enough that "navigation"—that is to say, in technical terms the finding of our exact position—could not be carried out except from the surface. On our own, with a few exceptions, we could tell only our depth and, in certain cases, the general direction in which we were traveling, as long as we had not yet reached the exact speed of the water (something that often took several hours) and the particles in suspension had a definite speed relative to our own. The surface vessel, on the contrary, can take its bearings either by the classical astronomical method or, if the skies are cloudy, by the "Loran"—to be more exact, in our case "Loran A" on the *Privateer* and "Loran A" or "Loran C" on the *Lynch.* The Loran system consists in establishing the difference, infinitesimal but measurable, between two simultaneous radio transmissions made by two different stations. System C is similar to system A but more modern and even more precise. Thus the *Privateer,* like every vessel in the world, could tell where it was with a good deal of precision, varying according to the condition of radio reception but generally to within a mile and often a half mile.

But the *Privateer* also had to follow us. To enable

it to do so we had four acoustic instruments aboard. First there were two "transponders" installed by Grumman. These devices remain silent except when interrogated by a signal: when the surface sends an acoustic "beep" in their direction, a signal transmitted by the water at a speed close to 1,500 meters a second, they acknowledge the knock by sending back a return signal. When this is received, the surface can deduce the distance and the direction of the mesoscaph.

The two other navigational devices had been installed by Navoceano. One is another "transponder," indicating only the distance; the other is a "transducer," which sends through the water every two seconds a sharp double "ping" at 4,000 hertz, and these signals enable the surface to determine our direction and depth. In practice this is the only beacon we use permanently. Every two seconds a condenser discharges and produces in the water this shrill sound which goes off to the surface with the latest news of position. We hear it distinctly wherever we are in the submarine, by day, by night, faithful and unfailing. It is one more link with the surface. The surface hears it continuously too. We hear it even when sleeping, eating, talking; we often hear it without hearing it; we hear it just the same. It could become obsessive, intolerable; it could drive us mad in half an hour. Actually it gives us pleasure, it has become a friend, a messenger, a perpetual comfort. We sometimes count its strokes for minutes at a time to make sure of its accuracy, as a doctor takes his patient's pulse. Our "pinger" always remained in good health; imperturbably it withstood water, sea, salt, pressure, temperature; nothing could make it fail in its duty. Every two seconds it struck, faithfully, faithfully, faithfully. After it had struck

1,296,000 times we could ascend and open the door to the sun.

During the whole afternoon and the first part of the night the *Ben Franklin* obediently kept gently rising. At 3:11 p.m. we had dumped another 15 kilograms of shot, and toward midnight, without anyone having adjusted the equilibrium in any way, the mesoscaph become stabilized at 200 meters and was at approximately the same temperature as the water outside.

30

THE LIFE IN THE SEA

Underwater life is not abundant in this region. Nevertheless, as soon as several of the searchlights are turned on, the plankton are unquestionably attracted. This positive phototropic reaction will permit us to make numerous interesting observations in the course of a month. The most common animals here are the salpas; unlike those I have seen many times in the Mediterranean these are not luminescent; they simply reflect the light of our searchlights, often brilliantly. They are curious little creatures resembling a small sack of transparent plastic, the size of a thimble or slightly larger, and containing minute dark-brown organs. In perpetual movement, inflating and deflating their bodies as a child blows up and empties a balloon, they proceed by describing circles, capering in the water, often making complete loops, frolicking like butterflies, which they strongly suggest when their shadows are thrown against the black background of the ocean. Some of them have a "tail" three or four times as long as their bodies, and others have none. Strangely enough, some tails occasionally meander through the water all by themselves. Erwin Aebersold witnessed an interesting spectacle: when a "tail" encounters a salpa it embraces it, caresses, cajoles, rolls itself around it like a scarf, a belt, a liana, and ends by attaching itself in the proper place, allowing the supposition that this is perhaps a spectacle "for adults only." Then the salpa resumes its transient life in the depths of the sea, capering, frolicking as though nothing had happened, but for it, at least, what an adventure—the adventure of its life, neither more nor less.

For us, too, what an adventure! For hours at a time we remain at the portholes observing these creatures, trying to predict or interpret their movements. Futilely, however. Their condition seems to be that of the most complete freedom, wherein lack of foresight can truly reign as absolute mistress. Free in the water, more so than birds in the air, for the salpa is not even affected by gravity: freer than the speck of dust in the wind, which ends by falling, the salpa has perhaps the freedom of a drop of vapor in a cloud. Yet it also has life, which endows it with fascination and mystery. How many hours have we spent watching these creatures, among the most primitive of beings, who have peopled the sea by millions since the dawn of time, and who perhaps beyond the primates and the starfish are our distant—very distant—ancestors.

At dawn on the third day the mesoscaph has

warmed up slightly, but it is still chilly—about 17° C. with a relative humidity of 72 percent.

When I wake up, at about 6 a.m., Kazimir is on watch and Frank Busby is working at his instruments in the stern of the mesoscaph. Erwin Aebersold and Chet May are sleeping. In agreement with Grumman it had been decided that at least two members of the crew would always be awake; in fact, a schedule of watches and work with extremely precise and rigorous timing, decreeing the minutes and hours reserved for rest, washing, eating, work, relaxation, and distractions, had been established. Our "life engineers" are anxious to keep us in good condition and at the same time to demonstrate the positive efficiency of the mission. They even wanted to inaugurate a day of twenty-two hours —instead of twenty-four—in order thus to gain "sixty hours" during the dive, that is, two and a half days. They gave that up when I pointed out that even at 600 meters one could still see traces of the light of day and it would be difficult to make the crew believe that this was the midnight sun.

I must report that, without any formal discussion whatever, without any dissent, without even a question, we paid no attention at all to the schedule thus established in advance. We arranged our own timetable according to circumstances, and we never had the slightest difficulty. The only rule we kept was that there must always be two awake as a safety measure. Someone on duty might have an accident, and it might occur when the mesoscaph was caught in a descending current or even drawn beyond its safe depth as the result of a leak, and then the "accident," regrettable in itself, could become a catastrophe.

So Kazimir is on watch when I get up. We have

breakfast together. It was not the best I have ever tasted. No hot croissants, no hot tea or hot chocolate in our underwater palace. A few cornflakes with powdered milk and a cup of tepid tea. That is all. We had, as I have mentioned, four tanks of hot water. We knew from the start that some of these tanks would not preserve their heat very well. Obviously these were the ones to use first, and indeed in two days the first one had cooled considerably.

We could, to be sure, have rummaged among the provisions aboard and found something else to complete this modest repast, but the knowledge that there was nothing hot to be had discouraged us from further search.

I shall return later to the question of food aboard. The perfect stability of the mesoscaph was enough to bring my thoughts back to this cheerier subject. For more than ten hours it had varied in depth less than its own height. This record was all the more important because never before had we tested its actual stability for more than a few hours. It had been carefully calculated, taking into account the relative compressibility of the water and of the hull, allowing for temperature and the compression of the oil and gas in each of the cells of the batteries and in the central reservoir. But the whole calculation was complex, and I must admit that the proof in practice, which we were in process of making, pleased me to the highest degree. Moreover it was an essential condition for the successful execution of our whole experiment.

31

THE DRIFT CONTINUES

At 9:32 a.m. on July 16 we feel, like millions of others, in complete communion with the three astronauts of Apollo 11 who were taking off for the most grandiose conquest ever undertaken by man. Thanks to the radio and the acoustic telephone, we are able to listen in on the countdown, and since I had had the good fortune to be present at previous launchings in the Apollo series, I was able fully to visualize the blast-off.

"Twelve, eleven, ten, nine, ignition sequence starts; six, five, three, two, one, zero. All engines running. Lift off, we have a lift off!"

We hear and surmise the rest, for the transmission is not of the best—the crowd shouting, "Go! Go!" to cheer on the rocket, the cries and tears of a million spectators drowned in the thunder of 180,000 horsepower charging at a gallop to the conquest of the moon.

At 10:00 a.m. there is 22.3 percent of oxygen in the air. It is a little too much; we close the cylinder for a few hours to reduce the excess.

Ken Haigh makes his first magnetic measurements with a proton magnetometer floating about 30 meters above the mesoscaph. This morning I observed the sea for a long time through one of the portholes. Without light there is nothing to see now. We are still oscillating between 196 and 200 meters. Everything is uniformly gray. With a searchlight, even a weak one (250 watts), there is planktonic life all about us. Here we see principally copepods, surrounded, however, by many other species of different sorts and forms. What interests me is to see that, if I fix my eyes on a speck of plankton, at the end of two or three minutes the speck disappears from my field of vision. Since we are moving with the speed of the water, this surprises me. At first I had hoped to be able to follow it with my eyes much longer. It seems that many factors interfere: the mesoscaph, though remarkably stable, often oscillates by a few meters, in a very slow rhythm to be sure. Sometimes these movements are produced merely by people moving about inside. But it oscillates horizontally, around itself. Our ship's compass shows us that the *Ben Franklin* has a tendency to proceed with the conning tower to the rear but that it also oscillates gently, turning to the right, to the left, at the will of imponderables that escape us, with a sort of horizontal pendulum movement of 30°, of 60°, of 100°; sometimes within several hours it makes a complete turn. These movements are almost imperceptible but along with the vertical oscillations and the weak but irregular movements of the water itself, the plankton does not remain stationary in front of our portholes but moves away gently, so gently, a few millimeters per second, and often less. Actually this has the advantage of bringing more species into our field of vision.

At 12:15 p.m. there is still 22 percent of oxygen in our atmosphere; rapid calculation shows us that we have used up about 0.3 liter per minute per man during two hours and fifteen minutes; this is normal and shows at least that our measuring apparatus is functioning properly.

We are off Fort Pierce. The surface of the sea is absolutely calm. "Sea state 0," the *Privateer* informs us. We picture the numerous private yachts fishing in this region which we know so well, hauling in swordfish, sailfish, blue marlin, even sharks. But where, after all, are all these famous game fish? We have been drifting for two days right here in their bailiwick and we have not yet seen a single one.

We are still motionless in relation to the water. Our stability is perfect, 203 meters, 200 meters, 202 meters. The humidity is bearable; it remains in the neighborhood of 70 to 75 percent. (We have enough silica gel.) On the other hand, we are not warm, and yet the thermometer reads 19°C.

Now it is night. At 10:15 p.m. I go to bed. Frank Busby will keep watch during the night, at the start with Don Kazimir, later with Erwin Aebersold. Frank is the night owl of the group. For a month he will sleep by day and work at night. Ken Haigh will do the opposite; so for a month one or the other will be on watch at all times and there will be continuous attention given to the oceanographic and especially the acoustic equipment.

Next morning, barely awake, I go to look at the

recording manometer: during the night the *Ben Franklin* has oscillated between 208 and 210 meters. One really could not demand greater precision. The other recording devices have functioned well too: for every two seconds, temperature of the water, depth, salinity, and speed of sound in water are registered on magnetic tapes. All is well aboard; morale is high, and how should it not be? We had hot coffee this morning. These six words may seem trivial or even the result of culpable materialism, but since our departure we have had nothing that was really hot. Two of our hot water tanks at least have defects in their thermal insulation and the temperature of the water has fallen appreciably, so this morning we decided to turn to one of the good tanks. But we economize on water to the maximum; we even catch in a little pail the cool water that comes out of the pipe before the hot water. To avoid reheating this pipe unnecessarily, we fill a thermos bottle each time with four or five cups of hot water; also whenever anyone takes a cup of tea or coffee he inquires all around if anyone else would like one too. This habit is easily acquired and results in our not losing a single drop and utilizing the calories to a maximum. The surrounding temperature of 19°C. still seems to us an absolute minimum: we have been in Florida so long that we are not used to the "cold."

A good exercise is changing the panels of lithium hydroxide, the substance that absorbs carbon dioxide from the air and which little by little—in three days according to our calculations—becomes saturated with it. It is a disagreeable job. As soon as one moves these panels, especially the replacement panels, they fill the air with their "powder" which irritates the throat and causes coughing. For five minutes the whole crew coughs so hard as to make you wonder whether they are not producing more carbon dioxide than the panels will ever absorb.

Toward midday, with Frank Busby, we try to get some idea of the color of the water; but the light is of such low intensity that the color is difficult to determine. The human eye sees by means of the rods and cones—photosensitive cells within the retina. The cones detect the difference in wavelengths—in other words they perceive color; these are the cells used in daylight. In the twilight the rods, incomparably more sensitive than the cones, are employed; these are bypassed when the light is relatively intense. The rods, however, do not distinguish color; that is why at night all cats are gray—and all fish as well. I had brought along a fine spectrum which Eastman Kodak Company published a short time before our departure, showing in large size the different colors with their corresponding wavelengths. We try to compare this spectrum with the color of the sea; one gets the impression that the light toward the top is on the order of 485 angstroms, and toward the middle 430, but at moments it suddenly seems more green than blue. If we were to descend farther, to a point where the light of day disappears, it would no longer be either green or violet; it would simply be gray.

At 11:00 a.m. the surface gave us our position, which we immediately entered on the chart; our speed is between 1.5 and 2 knots. At 2:00 p.m. we pass off Cape Kennedy. Suddenly Frank, who has recorded our position on his special charts of the Gulf Stream, cries: "There's something wrong! We can't be where the surface says we are. At this depth the temperature would be different."

In fact, off Florida, for a given depth the temperature of the water increases as one moves toward the east. At the latitude of Fort Pierce it changes at 200 meters depth from 6° to 20°C. over a distance of only 60 kilometers. We call the surface again and once more ask for our position: there has in fact been an error. Bravo for Frank! Navigation by thermometer, suggested by Franklin two hundred years ago, has proved itself. Probably the *Ben Franklin* has never deserved its name so well.

Early in the afternoon, one of the electric motors in the life-support system refused to stop on signal. Not one of us knows the details of its construction, but we have aboard the tools necessary, in theory at least, for any kind of repair. Ken Haigh and Chet May go to work; almost at once they find themselves stymied by a problem of disassembling. No matter. They call the surface: "Can you tell us which way to turn the lynch pin to get it out?" The surface has everything at its disposal: blueprints, electric circuit diagrams, and others; and what it does not have it can get immediately by calling on radio either Palm Beach or Bethpage, which are on the alert twenty-four hours a day. Promptly we have the answer. Work begins again. The repair takes a good four hours, and the system functions again as well as it did in the beginning.

This constant collaboration with the surface is an altogether new idea for me. When I used to dive in the *Trieste* we would barely have begun to descend when we were thrown completely on our own resources. Occasionally I had to make repairs aboard but I made them myself, and it never occurred to me to ask for any instruction from the surface, which, moreover, would have known nothing whatever about it. Here everything is more complex, beginning with the submarine itself, and the mass of information necessary—it would seem—exceeds the capacity of a single human brain, or even of six brains. To put it more accurately, it is a different technique; the same as that employed in space where the information is actually divided between the crew of the spacecraft and the crew on the ground. Since this method has been created, it is put to use even in cases where, at least to Europeans, it would seem unnecessary.

32

SECOND EXCURSION TO THE BOTTOM

We are making preparations for our second excursion to the bottom. Frank Busby requests that we arrive in the early evening.

"The bottom will be at 450 meters," he says, "and the temperature will be 7.15°C."

We shall see!

At 4.57 p.m. the maneuver begins. We start by orienting the bow of the mesoscaph toward the north so that at the moment we land on the bottom the guide rope will take on its function and we will drift bow forward. We also turn two of the motors vertically so that we can stop the descent immediately if there should be an obstacle or an accident.

At 5:05 we are at 198 meters. So we still have to descend some 250 meters and cool down from 17.30°C.

to 7.15°—that is, by 10.15°. To do this, in view of temperature and depth conditions, it will be necessary, according to calculations worked out on the basis of the data accumulated during the first dives, to increase our weight by some 350 pounds.

I decide to descend relatively slowly, at an average of 5 centimeters a second, to allow the temperature of the mesoscaph to adjust to that of the water; when we are on the bottom, balanced on the guide rope, we will gradually get rid of the excess water taken aboard and thus keep the submarine in equilibrium as it continues to grow colder.

Actually we do not have aboard a device that would permit us to carry out this program to within a meter or a kilogram—that is, a device for measuring the water as it enters or leaves the regulators. Without this instrument—we could not find a satisfactory one on the market—it is extremely difficult to know exactly how much water we take on or blow out, the volume of the water depending obviously not only on the depth but also on the variable size of the openings of the valves.

At 5:05, then, I open an intake valve for four seconds. The water comes hissing into the pipe, forced into the port VBT by a pressure of 20 kilograms per square centimeter.

At 5:10 still no reaction; once more I let in water for four seconds. Two minutes later comes the first reaction: very gently the mesoscaph begins to descend; it moves 3 meters in ten minutes. I let in water again for four seconds; this time we descend 8 meters in five minutes. All is going well; we can increase the speed a little. Seven more times during the next half hour I let in water, each time, according to circumstances, from ten to thirty seconds. At 6:00 p.m. we are at 365 meters and we are descending according to program at an average of 5 centimeters per second, one hundred thousand times slower than Apollo 11 is moving at that instant!

Through the porthole we see the water moving gently upward as we descend. At 6:30 we are at 448 meters, about 10 meters from the bottom. Behind our porthole is a spectacle to tempt a whale: playing in the beam of our searchlight millions of shrimp seem to be twisting and turning in all directions. In all probability these are "krill," the name given by the Norwegians to a planktonic species of tiny crustaceans, the euphausids. These are often luminescent; it is said that when they are concentrated on the surface of the sea a remarkable luminosity spreads over the entire surface. Here they are not luminous, but so plentiful that we could almost expect to see a blue whale sounding, one of those huge creatures that feed extensively on plankton and particularly on copepods and euphausids.

At 6:35 p.m. the guide rope performs its role perfectly and the mesoscaph comes to a halt all by itself without our having to discharge water or use the motors.

Depth 425 meters, water temperature 7.18°C. Frank Busby has correctly predicted the situation. There are moments when we wonder why he has come along, for he already knows the Gulf Stream so well. But, as we have seen, it is precisely because the Gulf Stream is relatively well known to oceanographers that it is worth studying more closely, to accumulate more data about it.

The Gulf Stream continues to show itself here, despite the depth; we are slowly drifting northward in conditions that are magnificent for observation. A few crabs, a few fish, and an anemone move past the portholes. The bottom is dotted with small holes, most of them empty, but in a few a crab or a fish is stationed. Too bad if a fish tries to enter one of these craters already occupied by a crab. I have just been able to observe one: he did not see the crab, which can be seen clearly from above, but at the last moment only a backward dart saved him from the claw which the crustacean thrust out in front of him like a crossbow; happily for the fish, unhappily for the crab—and for the next fish too, for the crab also has to have his food.

We call the surface on the telephone. Communication is difficult; there are terrific reverberations, echoes; you might think yourself in a great cathedral where the voice of the orator ricochets from every pillar and vault before reaching you.

"You are unreadable," we say.

A hopeless gargling scrambles the reply.

"Unreadable," we repeat.

"Readable, readable," the echo answers.

The *Privateer* comes back closer to the vertical and its voice becomes clearer.

The hull of the mesoscaph is growing colder; the interior temperature is now 16°C.; it will drop a good deal further during the next four hours. We have gradually grown heavier and are now resting on the bottom, motionless.

At 8:40 p.m. we decide to equalize the pressure in the port VBT in order to be able to expel the water and reascend a few meters. For this purpose I open a cylinder of compressed air or, to be more accurate, the pipe connected to that cylinder which itself is on the outside on the bridge of the mesoscaph. This time it is the air that hisses as it enters. I stop when the manometer indicates 50 atmospheres—that is, 5 more than the pressure of the sea. From then on it is enough to open the VBT in order to expel the water. In a few minutes we lose enough weight and the drift begins again at 10 to 12 meters above the bottom. The interior temperature drops another degree.

33

THE RUBICON

It is now exactly three days since the dive began. At the beginning of this expedition it was decided in principle that it should last either three days or a month: three days if major obstacles made us fear from the start that we would not be able to complete it, a month if all went well. It is now necessary to make an official decision. As a matter of fact, the decision has been growing steadily more obvious during these seventy-two hours. Everything aboard was going so well, the Navy's oceanographic instruments functioning even beyond what could have been hoped, that there was no reason to abandon the expedition; we were all in agreement about going on. At 10:00 p.m. Frank is on the telephone to tell Bill Rand that he proposes to continue the mission; Bill agrees on his part—the ocean is so fair, so magnificently calm, that there is no problem for him. He asks the opinions of Don Kazimir and myself. As a joke Frank replies that everyone is in agreement except Kazimir, the captain. The latter leaps to the telephone and corrects this: "I agree too!"

With a gurgling of bubbles the laughter from the surface reaches us at the speed of sound, 1490.7 meters per second, according to the Navy apparatus.

Then the acoustic explosions begin again, fifty times between 10.05 and 10.54, a rate of one a minute. It is now very cold (13°C.); our clothes are no longer adequate to the circumstances. They do pretty well for average temperatures when we are at about 200 meters, but as soon as we descend toward 400, 500, or 600 meters and remain there for some hours, generally motionless, the cold is pitilessly penetrating. To warm ourselves we do gymnastics or run in place, five minutes, ten minutes, fifteen minutes, but it must be admitted that this is not great fun. For this expedition, well and good; it is possible to bear up for a month, twenty-six days more. But if the mesoscaph were called upon to make a long dive in the cold waters of Alaska, for example, other arrangements would be necessary. First of all, the hull would have to be insulated, something we had thought of but had given up for various reasons. Also we would need insulated clothes, not of synthetic fiber but probably of goose down, which offers the best practical insulation available because of its extremely light weight. One might also have insulated clothes provided with electric wiring for warmth; to allow people to move around in the mesoscaph there could be electric outlets strategically placed all along the hull. However, these are plans for the future. For this first expedition the main thing is to hold out; it is the one key that will open the door to all further expeditions.

At 1:00 a.m. on July 18 Frank Busby has finished his observations and suggests that we ascend. For this, theoretically, it is only necessary to blow out the water taken aboard for descent. The mesoscaph will then be lighter, will check its ascent below the depth from which we started (200 meters) since it will still be cold, and from there will gradually rise to 200 meters as it warms up again. This once more is theory; what happens in practice remains to be seen.

First I turn on the air in the VBT for forty seconds. At this point the regulator seems to be empty; through the porthole at the pilot's post air can be seen coming out. For a moment we look at each other, Erwin Aebersold and I; it is impossible that these forty seconds would be sufficient to expel the water that we took aboard in order to descend. Then where is that water? Has a slow leak in one of the valves already driven out the water without our noticing? No, for then we would already have risen. It was Aebersold who designed the details of the variable-ballast-tank system; he makes a quick sketch, glances at the pendulum, and decides that we are tilted too far forward to empty the VBT completely; we have to add weight to the stern so that the entry tube for the air is effectively submerged in the water that is to be driven out. For this purpose we have trim tanks. An electric commutator is engaged; the mesoscaph rights itself again but not quite enough, for the whole crew at this moment is in the bow. For the final adjustment four of us go to the stern; we are now tilted more than enough. The water goes out once more and the mesoscaph ascends gradually as we wish it to do: 100 meters in a little more than an hour, then another 100 meters in three hours, even more slowly, as we approach the state of equilibrium. Erwin Aebersold relieves me at the pilot's station.

To avoid this problem the VBTs are usually placed in the center of the submarine. In our case this arrangement was not possible; they would have taken up too much room in the interior of the hull and it had been necessary, therefore, to place the VBTs on the outside of the hull along its entire length.

Toward 3:00 a.m. I take up a position at a porthole in the stern of the mesoscaph. An exterior floodlight is on. I have two cameras within reach in case of need; on my left a Paillard for 16-millimeter movies and on my right a Minolta for stills. For a long while, wrapped in a blanket—it is now about 10°C.—I look through the porthole. Since there is not much to see, at 5:45 a.m. I go to bed, hoping for a few hours sleep.

34

THE ATTACK

It was while I was asleep—restlessly, because of the cold, the inevitable noise, and perhaps excessive fatigue as well—that the attack took place.

This event seemed to me so important that I wanted immediately to commit every possible detail to paper. At the moment it occurred, Ken Haigh, Frank Busby, and Erwin Aebersold were awake, so I brought them together and tried to get some agreement about the size of the aggressors, their characteristics, and the circumstances of the attack. It is very important to note such impressions promptly, for they change readily, even with the best will in the world and the greatest sincerity and objectivity. The way hunting and fishing stories proliferate and expand is well known. I determined to have this account utterly objective.

The incident occurred at 6:09 a.m. at a depth of 252 meters. No one was at the portholes. An exterior floodlight had been on all night. Frank Busby, working in the section of the stern that we called the laboratory, suddenly saw something glinting through the porthole. He rushed to the window and saw a magnificent swordfish of the kind that is known to live in the depths of the Gulf Stream and is among the handsomest trophies of the Florida fishermen. Frank had time to see that it was unquestionably a broadbill—that is, it had a large "beak" or "nose," unlike the marlin's which is rounded and narrower. The function of the sword is, it is thought, to impale fish the swordfish intends to eat, perhaps holding them in reserve for a few minutes while continuing to hunt and eating them eventually in a protected spot secure from possible rivals. Some swordfish exceed 4 meters in length: in the Gulf Stream fishermen often catch specimens 2 or 3 meters long. The one in question, according to Frank, could not have been more than 1.5 to 1.8 meters, but length is difficult to estimate exactly. Frank observes the swordfish for a few seconds, during which it seems very agitated and comes to take a close look into the porthole through which it itself is being observed. It appears to parade up and down in front of the window, dashing to and fro, not knowing how to interpret our presence, swimming a few meters and then returning, as though fascinated by our great Plexiglas eye. Suddenly it attacks, dashing straight forward and striking the hull of the mesoscaph with the point of the sword, aiming perhaps for the porthole but only hitting the steel of the hull 15 to 20 centimeters beneath the Plexiglas. Frank and Ken Haigh, not far from the porthole, both hear the impact clearly. Frank rushes to the bow of the mesoscaph to get his camera. When he returns the attacker has disappeared, but meanwhile Ken Haigh distinctly sees a second swordfish which, as though directing the attack from a distance, remains at a prudent remove, at the limit of our searchlight range. The whole incident lasted perhaps a minute. Erwin Aebersold, who was breakfasting in the bow of the mesoscaph, did not hear the blow and had seen nothing through the forward porthole.

Listening to this account, I thought of my two cameras which I had had ready at hand an hour before and with which I could have recorded so many details now gone forever. Forever? Who knows? Perhaps the incident will be repeated. Another attack of this sort—and this is not the least interesting part—had been observed in similar circumstances. The *Alvin,* a small submarine mostly designed by the engineer Al Vine of Woods Hole, had been the object of such an attack during the course of one of its numerous dives. That swordfish, less fortunate, had remained wedged in the submarine, his sword caught in the plastic and fiberglass superstructure. The pilot, not knowing whether there had been serious damage, decided to ascend; thus he brought back the fish so that it could be observed on the surface, taken to the harbor, shown to the Woods Hole zoologists, and eaten that same evening.

Why do these swordfish attack submarines? Are they fascinated, hypnotized by the portholes? If they mistake submarines for monsters, which perhaps they actually are, what courage these fishes show in attempting to impale an adversary so much bigger than themselves. Decidedly the psychology of fishes is barely understood; there is an immense and fascinating field here, but difficult to study. It can only be hoped that new observations of this kind will be made.

Does an attack like this present any danger for a submarine? To tell the truth, I do not think so, but one can imagine extreme cases in which the situation might become critical. If the submarine, as a result of any cause whatever (errors of calculation, pilot error, enemy action), is at the point of rupture, the violent shock caused by a creature of this size striking at a particular point at a speed of some 10 meters a second, might make just the crucial difference. These are extreme conditions, highly improbable. On the other hand, after the attack on the *Alvin,* specialists (in what field, I do not know) calculated that if the sword had struck at a certain angle exactly at the ring that holds the porthole, ring and porthole could have been removed, causing the destruction of the submarine (and an insoluble problem for investigators even if the vessel were recovered). As for our portholes, each is held in place by a steel ring secured by a sufficient number of high-resistance screws, so I consider that the hypothesis of such a rupture can safely be discarded. Obviously if a porthole were destroyed, the situation aboard would not be comfortable. At a depth of 250 meters the water would rush into the hull at a speed of some 70 meters a second and with a flow exceeding 1 cubic meter a second. In a few moments more water would have entered than could be compensated for with the emergency ballast and the various ballast tanks, and the overweighted hull would rapidly go to the bottom with no chance for its passengers to escape.

All goes well now aboard the *Ben Franklin* except several of us have caught severe colds. Don Kazimir gets out a small suitcase and proudly opens it in front of us. Everyone admires the innumerable little bottles it contains, pills of every kind and color. No doubt blue for headache, yellow for earache, green to support morale. He picks out a bottle of red tablets, excellent for colds, according to the list of instructions glued to the inside of the cover. Personally, I do not greatly appreciate this kind of medication. It may be good for alleviating symptoms but it certainly does not do as much to cure a cold as would a nice walk along the beach. However, in our case a walk is impossible, and even if the secondary effects of the pills are unknown, it is better to get rid of the colds in a hurry, rather than risk a real epidemic.

In the course of the afternoon we can see clearly that the current is going past us. This seems impossible, since nothing is holding us back in the water. But the reason for it lies in a phenomenon that occurs also with free balloons. In a free balloon carried by the wind, which theoretically one does not feel, however strong it may be, a zero velocity is maintained in relation to the air. Nevertheless, as once happened to me during a trip I made with my cousin Donald Piccard in Texas, the wind suddenly can be felt distinctly and the balloon may be shaken by eddies; this usually occurs if the balloon suddenly changes altitude and lasts until it adjusts to the new velocity of the air. It can also happen if the balloon suddenly enters an ascending or descending current. The mesoscaph, actually, was simply moving slower than the water. Since underwater phenomena are generally very gradual and since the inertia of the mesoscaph is considerable, effects are felt for quite a while and often belatedly. Because the speed of the current diminishes with depth, the result is that the water at the top of the mesoscaph moves faster than at the level of its keel; this "shearing" effect of the water can easily amount to 2 to 3 centimeters a second for the height of the mesoscaph, where, for example, the speed of the Gulf Stream is 3 knots on the surface and 2 knots at a depth of 200 meters. Between the keel and the portholes the breaking effect could thus easily amount to 1 centimeter per second, which could readily be observed by watching the movement of the plankton. Now this value is confirmed by the Navy current meter which suddenly indicates a speed of 0.05 knot in relation to the water.

What poses a more complex problem is that this same current meter indicates we are all of a sudden drifting toward the south. Until there is proof to the contrary we concede that we are probably on the edge of a secondary eddy whose center is continuing to advance toward the north. Our speed, however, and this gives us some small concern, is certainly below what we had expected, averaging only 1.3 knots since the start.

35

PLANKTON

In observing the water diligently we have the good luck —though it is not luck at all, since the construction of the mesoscaph and the whole expedition have been centered on this problem—to be able to see in extremely favorable conditions an enormous quantity of plankton. When I say "enormous" I should make it clear that I am taking into account the entire duration of our dive. Rarely were we submerged in plankton, but by spending the hours we did at the portholes we saw thousands of different minuscule forms in the water, scores of which soon became familiar to us. For example, I have often seen a sort of miniature train drawn by a comparably tiny steam locomotive, all of it 2 to 3 centimeters at most. None of us could identify it; it is obviously some planktonic form, but is it a larva in process of development or an adult creature? Often the larvae of crustaceans have aspects in an early stage so totally different from their adult appearance that it takes an experienced marine biologist to identify them.

As soon as we turn on our floodlights the plankton begins to gather in front of our portholes. At the end of an hour, then, there is a veritable mass, a teeming crowd constituting a spectacle of the greatest interest. There are, of course, the euphausids already mentioned; there are the true, large shrimp; there are other creatures such as the *Sagitta* (arrowworms), tiny half-transparent rods that can be seen for long moments completely immobile and which then suddenly fly like arrows (hence their name) for a distance of 20 to 30 centimeters. Biologists say that they throw themselves thus upon their prey—sometimes as big as themselves—but I have never been able to see them capture anything whatever: their movements are too rapid. Another of our little planktonic friends which I have often observed is the *Phronima.* It is actually a small shrimp and is found in two aspects: sometimes solitary in the water, sometimes inside a sort of transparent, plastic envelope that is simply the tunic of one of the *Phronima's* prey. In fact, the *Phronima* frequently attacks the tunicates, eating out the interior and keeping the transparent carapace for putting its eggs in to hatch. Often we have seen it half inside the "shell," its feet extending from the back so as to permit it to swim more vigorously. The British naturalist Richard Carrington has compared this with a conscientious mother pushing a pram filled with her offspring. The favorite prey *Phronima* uses for this purpose is *Pyrosoma,* a variety of tunicate that lives in colonies. We have seen a myriad of *Pyrosoma,* but the envelopes used by the

Phronima within the scope of our observations seem rather to have come from salpas.

I do not know how biologists have managed to snatch so many secrets from the depths of the ocean without either bathyscaph or mesoscaph. Obviously one cannot establish a general rule on the basis of only a few examples, but direct observation is incomparably richer than any that can be made otherwise. The phenomenon of bioluminescence is striking in this respect; *Pyrosoma* owes its name ("fire body") to its supposed bioluminescence. We saw whole processions of these animals time after time during our expedition but not one of them was phosphorescent. Is this because the water that surrounded us was particularly calm? Is it a question of the season? Of the depth? Is the pyrosome luminous only at special times?

Here the bulk of the plankton is undoubtedly made up mainly of copepods, bizarre crustaceans which are said to be able to attain a length of as much as 1 centimeter but which, judging by what we have seen, rarely exceed 2 to 3 millimeters. These copepods —and there exist innumerable varieties adapted to fresh water as well as salt and to the water of marshes and caverns—serve as the basic food of a great variety of animals, particularly whales, herring, and sea birds. Like the *Sagitta,* they often remain motionless for seconds at a time, then suddenly dart off straight in one direction and stop as abruptly as they started. It seems, though we were not able to observe it directly, that they sometimes move about carrying in one of their numerous small feet a "bottle" or spermatophore, which is destined for a female copepod. When the mate of their choice has accepted—generally not without a show of reluctance—this marriage pledge, the male copepod goes off in search of adventure, for the bride finds herself supplied by this one event for the rest of her life.

The most beautiful spectacle of planktonic life we saw is certainly that afforded by the long chains of salpas. Individual salpas, which were described in Chapter 30 and which I have also seen in the Mediterranean, are usually about 2 or 3 centimeters long, although sometimes they have lengths of 10 to 15 centimeters and diameters of 1 to 3 centimeters. However, in the Gulf Stream these creatures also appear in a totally different form: arranged in long chains and undulating in the sea like a serpent in the grass—even more freely, since here they revel in the three dimensions of space. They approach the beam of light, which is odd, for they are seeing without seeing, since they have no eyes and all obviously obey the same rudimentary brain, but they are certainly endowed with a will, or at least a specific tendency. They undulate wonderfully, wandering around the mesoscaph, passing and repassing in front of the portholes and offering us the incomparable spectacle of the miraculous details of their bodies. Each individual powerfully reflects, from one or two mirrorlike points, the light that falls on it. The closer they come to us the more details we can see and the more clearly we realize that these details are infinite. These chains may be several meters long, very often 3 to 4, sometimes 5 to 6 or even much longer. At times, weary of undulating, they roll up in a spiral and let themselves drift idly like nebulae in space. Sometimes they form a necklace, sometimes a bracelet, sometimes a garland, always something beautiful and marvelously delicate. Quite probably these chains are simply an intermediate stage in their development.

The life of these salpas is strange. In their individual state they are probably sexual. The second generation is not, but on their tentacles buds develop which grow and become progressively recognizable as salpas too. Arrived at maturity, these creatures, in which the two sexes are once more represented, detach themselves from the mother tail, the chain breaks, and the cycle begins over again.

What beauty and what delicacy! Seen in the beam of our floodlights, they are jewels that Benevenuto Cellini would not have disdained: what profusion of tiny flourishes, what tracery, what decorations, what minutiae of elaboration. We often watched them pass slowly in front of our portholes: they could easily be taken for jewels of solid red gold set in silver or platinum with pearls of luminous purity. If we should try to take them in our hands they would slip between our fingers and disappear. The only remnant on our palms would be a bit of that sticky jelly which constitutes, for biologists at least, the entire difference between life and lifeless matter. On many occasions I have seen what was probably an adult salpa after it had abandoned its progeny: it appeared to be a big individual salpa but with a whole series of long filaments trailing behind. It reminded me of a dirigible landing at night in the beam of floodlights with its ropes already lowered so that the ground crew can secure it. What mysteries and what grandeur, just as there are mysteries everywhere else, to be sure. Here we have time to look, to observe, to think.

36

AMBIENCE

This evening I go to bed early. The day fades fast at this depth. However, it is not easy to sleep. For one thing, there is so much to think about, and then a few centimeters from my eyes I always have the porthole "wide open" to the sea. Even if there are very few *Noctiluca,* I often see luminous points pass my window and I am always afraid if I close my eyes I may miss an important spectacle, like that of the swordfish. Also there is noise, especially in the evening when almost everyone is up. Since the light on board seems the same day or night, those who sleep during the day do not feel that it is night when they are at work, and they are not yet really trained. As for those who work during the day, are they any better off? Certainly not always. We still have to learn more respect for the sleep of others, whatever the hours. For example, we have a cassette player aboard. Everyone has brought several cassettes. On the first day, with Mozart, Rossini, "Dr. Zhivago," and "The Yellow Submarine" we think we have a large selection to satisfy all tastes, all desires. On the second day there are still "The Yellow Submarine," "Dr. Zhivago," Rossini, and Mozart; on the third day one can still vary the order but on the fourth and the fifth it begins to be tiresome. We still have twenty-five days ahead of us. Along with the player we have a set of earphones of excellent quality. This occasionally permits one or another of us to isolate himself with Mozart (or the Beatles), but more often the concert is imposed on us while we try to sleep or to listen to the reverberations from the explosions set off by the *Privateer,* or to understand the life cycle of the *Phronima* nursemaids. On another expedition each of us will have to have a pair of earphones, and the mesoscaph will have to be wired so that a crew member can plug in wherever he happens to be. I finally go to sleep, leaving the submarine in the experienced hands of Erwin and Kaz.

Next day, July 19, we are in a good state of equilibrium about 220 meters down. Inside the mesoscaph it is not quite 18°C.; outside, 16.5°C. When in equilibrium the interior temperature is usually 1° to 1.5° above the temperature outside.

The drift continues northward at slow speed, 1.6 knots on the average since our departure from off Palm Beach. We spent this last night off Daytona Beach, that coast traversed by the famous track of the automobile races, where one sees as many cars as bathers. To combat the cold and relative physical inactivity, we go on running in place. Chet May has brought with him a rubber-tired wheel some 20 centimeters in diameter, equipped with a pair of handles on which one supports himself while rolling back and forth to strengthen the back muscles—a true torture on the wheel in the age of submarine exploration. After "breakfast" we turn out all the interior lights of the mesoscaph and remain for a while in "obscurity" to enjoy the direct light of day. At the end of ten minutes one can see the interior of the mesoscaph fairly well; one can move about and find one's way, avoid running into things while crossing the corridor, at least running into no more than when the light is on (for the corridor remains an unalterable 80 centimeters wide), but it is not bright enough to read a book.

37

A MONTH—BUT A MONTH FROM WHEN?

With lights turned on again we find ourselves assembled in the bow of the mesoscaph. Without planning to do so, we fall to discussing the duration of the mission. We all agree to remain aboard, according to program, for a month's time. But to know when the month ends it is first necessary to know what a month implies; for example, in number of hours and when the month is said to have begun. We are all agreed that in this case a month means thirty days—that is, 720 hours. However, there are several views about the beginning of the dive.

If you talk about a month's drift, the expedition began at the moment when the hawser from the tug was dropped, that is, July 14 at 8:25 p.m.

If you talk about a month's expedition, the moment of departure is when the whole crew was aboard, about 8:30.

If you talk about a month of isolation, it is 8:34 that should be considered, the instant when the hatch was closed. One might also consider 8:40, when the flood valves were opened—that is, the beginning of the dive—and as a last possibility, 8:54, the moment when the mesoscaph was totally submerged.

Don Kazimir inclines toward 8:40; Frank Busby and I toward 8:54.

As a matter of fact, I shall recommend that we remain submerged until the morning of August 15 so as not to have to surface during the night, which might be difficult, especially if the sea is not perfectly calm.

This discussion may appear somewhat pointless; I report it nevertheless. In reality it was amusing; it reflected one of the preoccupations to which we had a right, for completely enthusiastic though we were, we paid heed to the stages of this dive, including its end, with an interest that I still consider completely justified.

Toward the end of the afternoon we notice that the mesoscaph has a tendency to move westward; since the surface informs us that we are still in the center of the Gulf Stream, this hardly causes us any concern.

It is time to change the panels of lithium hydroxide. Before doing so we measure the percentage of carbon dioxide in our atmosphere: Erwin Aebersold finds 1.4 percent. This figure is lower than our offhand estimates, and rather quickly we decide that we ought to repeat the measurement until the apparatus gives a proper reading. A second measurement indicates 1.2 percent (still less), a third reading 1.4 percent again. I must say that the apparatus in question has never inspired me with confidence; these results are decidedly too capricious.

Don Kazimir tells me that he has always used it successfully in the Navy, but since he has never compared it with any other it is hard to judge its accuracy. It hardly matters, for we know very well, Kazimir and I, and even Busby and Haigh, all of us acquainted with submarines, that as long as you get no headaches there is no danger of excessive carbon dioxide. It is curious that the body has been thus provided with a warning system against a danger that rarely exists in nature. The other "alarm systems" that warn us against fire, excessive fatigue, poison, correspond to phenomena and risks that cavemen might have encountered at any time. Perhaps this one developed as a protection from the natural carbon dioxide that is occasionally found in caves.

Chet May continues his microbe hunting. Every day he collects specimens, puts them in "culture broths," and observes the development of the colonies he is able to isolate.

There is something peculiar in this research: we know that a situation might occur in which these bacteria would present a definite danger to us. It is for this reason that Chet surveys them with so attentive an eye. At the same time he has the feeling of not doing his job if he fails to find them, and so he is caught, and we with him, in a double game, the sort of paradox that makes us rejoice with him when he finds them, when he isolates them, when he makes them thrive in his flasks. Also we fear more than anything else that these wicked little creatures might oblige us to interrupt our marvelous voyage. When I ask Chet May how "his" bacteria are doing, he replies with an air of mystery: "Pretty good so far."

At the end of the afternoon I try to trap some specimens with the plankton tube. There is a story connected with this plankton tube.

38

CAPTAIN NIC

In 1965, well before my association with Grumman, I had had a telephone call from a kind of impresario, one "Captain Nic," who invited me to take part in the production of a film on oceanography to be made in Florida. I was in Switzerland at the time but I accepted nevertheless, and a few days later I found myself with a group in Saint Lucie, Florida, one of those communities so characteristic of that state: around a country club and its golf course and enormous swimming pool are some hundreds of small houses of one, two, or three rooms, built along tidy streets that run through the grounds in concentric circles. Nic, the impresario, his family, his assistants, his director, wearing huge sunglasses, his cameramen, some technical advisers biological and otherwise—in short the whole traditional general staff of the film world—were assembled there, ready to set the cameras rolling.

The program under consideration was excellent but unfortunately there was no key subject to connect the different episodes. I proposed the Gulf Stream, about which I had spoken at the National Science Foundation in Washington the year before. The idea was accepted. We left for the Gulf Stream in a rented boat to find out about the current. A small submarine was also rented from a friend, John Perry, who from the start was named technical adviser along with Ed Link, the industrialist who gave his name to the flight simulator that has saved the lives of so many Allied pilots during and since World War Two.

A first meeting was arranged to study more closely the scientific possibilities of the Gulf Stream project. Various oceanographers took part, including Dr. Paul Fye, director of the Woods Hole Oceanographic Institution, Dr. L. V. Worthington, also of Woods Hole, Bill Richardson of Nova University at Fort Lauderdale, Florida, and Feenan Jenning of the Office of Naval Research in Washington. In the course of this conference devoted to the general outlines of the project, the possibility of drifting was recognized in the plans.

Dr. Fye proposed to include among the oceanographic equipment a water-resistant steel tube open to the sea at both ends and passing through the interior of the hull. In the interior portion there could be a transparent section of glass, a small pump to drive the sea water through the tube, and a pair of valves permitting the glassed-in portion to be isolated. With the pump one could bring water containing plankton inside the mesoscaph, isolate the plankton, and observe it at leisure while it was still under the pressure of the surrounding sea water and at the same temperature too, at least during the first few minutes.

The possibility of observing living and imprisoned plankton in these conditions was extremely tempting. There would, in fact, be an enormous difference between these living specimens and the miserable, lifeless forms crushed in nets and traps that are brought up at random into the traditional oceanographic boats. Dr. Fye went so far as to tell me that the whole expedition would be "profitable" if a single specimen of plankton could be examined under these conditions, quite apart from any other possible results.

Nic's film was never finished. His whole group disbanded one fine day, leading to the supposition that the tripods of his cameras had no foundation but the sand. I heard no more of him. There are often groups that get together in this way, hoping to make a fortune from an idea, then dissolve when the initial capital is eaten up, and drift off elsewhere.

Nevertheless I have remained grateful to Nic for the preparatory work he made it possible for me to accomplish within the framework of the Gulf Stream expedition. Moreover, there had been the scientific conference on the Gulf Stream, the idea had been accepted by recognized oceanographers, and there also remained the specific idea of the plankton tube. The tube was built and installed aboard the *Ben Franklin,* and although it did not figure directly in the principal program of the expedition, for we had no biologist aboard, it was understood that I would use it from time to time, chiefly to establish its effectiveness and, if necessary, to suggest improvements.

On this particular day, then, I tried it out for the

first time. To tell the truth, I did not catch anything and succeeded only in confirming some of my comrades in their conviction that the tube would never work. As to that, there is a sequel.

39

THE SEA IN MOTION

Something new tonight. I have had trouble falling asleep. I was a little concerned by the line continuously traced on our manometer showing our depth: indisputably we now have a certain tendency to rise—not very fast, but perceptibly. We began the day at 230 meters; at 8:00 a.m. we were at 208 meters; at noon 195 meters; at 6:00 p.m. at 183 meters; at 10:00 p.m. 181 meters. There were several theoretical explanations for this. We could be slowly losing compressed air through some of the valves, and this air could be accumulating either in the VBT or the ballast tank, thus gradually lightening the mesoscaph. Or perhaps the current itself was slowly rising. To assure ourselves that we were not accumulating air in the ballast tanks, we opened the flood valves several times, thus permitting any air that might be there to escape. But this—fortunately in one sense—had no effect, and so probably, though we could not then know with certainty, the current itself was rising slowly. How far would it carry us? As far as the surface if we let it? Would it descend again, or would it stop at a given level?

There is no way of making an assumption. Don Kazimir and Frank Busby are on the telephone with the surface. If the current is really rising, should we perhaps proceed with a "correction of trajectory" by utilizing our propulsion system to change our position in the current and find another "vein"? Finally from my berth I hear them agree to wait until tomorrow before deciding. In any case there is nothing urgent, but it will be necessary to keep an eye on the needles of our manometers.

Next morning, July 20, I wake up at 7 o'clock. It is comfortable in the mesoscaph. The water outside is about 17°C. A glance at the manometer reassures me; we are now at 197 meters; so we have descended somewhat. I leaf through the ship's log to discover the important facts of the night: 4:50 a.m., Kazimir had to take water into one of the VBTs, for the *Ben Franklin* had risen to 140 meters. This is why we are no higher now. The problem has simply been postponed; we will have to give it close attention. The mesoscaph is now oscillating around 200 meters. The graph shows that we are oscillating with an amplitude of about 10 meters and a frequency varying between fifteen and twenty minutes. Each time we rise we wonder whether it is a "true" rise, with the risk of going too high; but each time, without any action on our part, the mesoscaph descends again, to repeat its ascent a quarter of an hour later. We are now oscillating like a pendulum, quite regularly. It is probable that we are slightly heavier than the water at this moment and that our downward tendency is just compensated for by the slowly rising current. In any case, we are in a zone characterized by undersea waves, the so-called internal waves.

40

OUR FIRST SUNDAY UNDER WATER

What shall we do to celebrate this day? Obviously our regular routine, observations, and work must go on. Up above on the surface the *Privateer* keeps guard. Much farther off, at a distance of some 380,000 kilometers, Apollo 11 continues its flight, the lunar module is going to land on the moon, Neil Armstrong and Edwin Aldrin will take possession of a new world in the name of their country and of all mankind; down here as well work continues.

Originally we had thought that one of us (the captain?—the leader of the mission?) would conduct a kind of service or preside at a period of meditation. It happens that the six of us aboard do not all have the same philosophical and religious outlook. One of the six is against all forms of religion, to the extent that he bristles and his blood stops circulating (so it would seem) when someone says "God bless you" when he sneezes. Since mutual understanding aboard is essential, and in our case religion has no need of external forms, we have decided simply that on Sunday a Bible will be put at the disposal of those want it.

During the afternoon—*Rigoletto,* courtesy of the cassette player—I spend a long time at the portholes. The emptiness of the sea, its sky-gray color, prompt me to think of the incredible epic that is being written on the moon at this moment. The landing is supposed to take place at 2:23 p.m., our time. But the surface does not even mention it. They simply say that we are veering once more slightly toward the east and that probably no correction of trajectory will be necessary. Perhaps it is for this reason too that we have stopped our gradual ascent. But if the water becomes stable we shall be too heavy by the amount of the water taken on board last night. No matter, we can always get rid of it by blowing compressed air into the VBTs.

41

CHLOROPHYLL AND MINERALS

Toward 4 p.m., while waiting for news from the moon, I make some measurements of the chlorophyll and mineral content of the ocean with an apparatus built for us by W. G. Egan, a researcher in Grumman's Geo-Astrophysics section.

The importance of chlorophyll in the ocean, occurring in the phytoplankton, is obvious. It is the phytoplankton which, in the zone where enough sunlight still penetrates—let us say to a depth of no more than 150 to 200 meters—uses the energy of the sun as do the plants on earth to carry on the basic reaction of creating sugars or glucides from water (plenty of it is present) and the carbonic acid contained in it. This phytoplankton is later eaten by microzooplankton, then by zooplankton of the ordinary kind, which can also feed directly on the phytoplankton, It later serves as the basic food for the smaller fishes, notably herring, which in their turn are food for fish of average size, such as mackerel. The mackerel serve as food for the tuna, which are a source of food for the sperm whale. Since nature does not like overly rigid arrangements, it has also allowed the largest animals on earth, the blue whales, to feed directly —as already described—on zooplankton, the krill, masses of planktonic shrimp.

Man does not consume phytoplankton directly, any more than he eats grass or hay.

It takes a ton of grass to make 100 kilograms of beef, which are transformed finally into only a single kilogram (1 percent) of human weight. At sea, according to the estimates of the oceanographers, 1 ton of phytoplankton produces 100 kilograms of zooplankton, which will produce 10 kilograms of small fish, and these 10 kilograms will become 1 kilogram of big fish. However, of this final kilogram—of salmon, for instance—only 10 percent, that is 100 grams, is directly assimilated by the man who eats the fish. In the eyes of nature, therefore, it is ten times more advantageous to eat herring than salmon.

To return to the Egan apparatus for measuring the chlorophyll content of the ocean—that is, essentially the amount of phytoplankton—it happens that chlorophyll becomes fluorescent or, more accurately, luminescent in a beam of ultraviolet light with a wavelength of 3,660 angstroms, that is, 0.366 micron. Now we have outside the mesoscaph a bulb (in fact, two bulbs) producing a beam of this wavelength, a beam that falls on a special photoelectric cell after having traversed a small portion of sea. This cell reacts differently depending on the quantity of chlorophyll contained in the water through which it passes. The final determination of the amount is made from a special voltmeter. Considering that on the whole there is little phytoplankton in the Gulf Stream, the slightest variation in these measurements may be significant and useful.

The same apparatus can also throw a beam of light of 2,537 angstroms toward the photoelectric cell. Chlorophyll, to be sure, will no longer react, but it happens that a whole series of minerals will become luminescent; the results of these measurements will thus give us valuable information about the mineral riches of the water and the nature of the sea. To gain a general idea of this problem in the Gulf Stream I count on taking these readings three times a day.

Finally, the apparatus was supposed to count the naturally fluorescent particles—that is, to estimate the bioluminescence in the Gulf Stream. Unfortunately the comptometer did not function, and this kind of measurement will have to wait for a later occasion.

At 4:20 p.m. the surface finally sends this brief message, so eagerly awaited: "Apollo 11 has landed on the moon." Nothing more. One can only assume that there has been complete success. Decidedly the surface is not talkative. Also it seems that the ship-to-shore radio leaves much to be desired.

At 4:36 another brief message transmitted by Bruce Sorensen: "Two Americans have landed on the moon."

"Roger."

"Out."

That is all we shall know, today and for practically the whole duration of our dive, about the most awe-inspiring expedition ever accomplished by man.

42

NEW DESCENTS TO THE BOTTOM

The afternoon has been calm, but during the evening and night we have again oscillated considerably. Offhand one might think that it is the mesoscaph which, being in good balance, oscillates like a pendulum. Not so; it is simply internal waves which now carry us with decreased amplitude, generally between 5 and 10 meters. This corresponds, moreover, to a slight increase in our speed, as if the arm of the Gulf Stream where we are at present has to find its way by oscillating in the great mass of surrounding waters.

We are drifting quite consistently, following a line about 20 kilometers east from the 200-meter level line depth, as though the Gulf Stream were canalized there by the continental shelf. If this drift continues we shall have no trouble passing Cape Hatteras, which is a sort of Cape of Good Hope in the middle of our mission; we have the feeling that if we pass Hatteras without difficulty we will at least have accomplished the greater part of our mission. Beyond that, we expect to come to a much more tumultuous area, and we are afraid we may have to ascend to the surface several times and be towed back into the current. As far as Hatteras (latitude 35° 13′ N.) I do not anticipate any serious problems.

July 21: a national holiday in Belgium, the country that made it possible for Dr. Auguste Piccard to carry out his ascensions into the stratosphere and to build the first bathyscaph. What a distance we have come since then, and what a fine scientific investment for Belgium those first beginnings were.

This afternoon we are going down to the bottom again. If we find there a current that will allow us to cover some distance and observe an extended area, we will remain there, according to program, for twenty-four hours. The temperature of the water will be about 8°C. At the end of some ten hours the interior temperature of the mesoscaph will be in the neighborhood of 10°C., and then we will stay there ten to fourteen more hours, not counting some ten hours after the beginning of the ascent, before the interior temperature returns to a supportable range. In plain language, we have twenty-four polar hours ahead of us. Kaz folds his hands and prays that there will be no current on the bottom. In that case we will be able to finish our observations in two or three hours and our atmosphere will not have time to grow too cold.

At the end of the morning we are oscillating from 20 to 30 meters. Before beginning to descend Ken Haigh makes another series of acoustic measurements with the *Lynch;* he records the echoes on magnetic tape, and I see the strange waves march past on the oscilloscope. Midway between the mesoscaph and the surface "something" can be seen, hardly distinguishable. Ken is unable to interpret this echo immediately. Is it finally the mysterious "Deep Scattering Layer" that we so much hope to see, is it simply a thermocline—that is, a zone of abrupt temperature transition? In any case it is not our job to analyze the results yet. These will be studied ashore over a period of many months by Ken Haigh and other specialists.

The actual measurement is performed by setting off at a depth of 10 meters under the surface charges producing very "white"sounds—that is, a mixture of all the "acoustic colors," which in turn means of all the frequencies—and then measuring the difference between the waves reaching us directly from the *Lynch,* about 1.6 kilometers away, and those that come echoing from the bottom, which is some 150 meters below us.

At 2:00 p.m. we prepare to descend, encouraged, despite the prospect of the cold that awaits us, by good news from the moon: very briefly, as usual, we are told that the lunar module has successfully taken off from the moon and is in lunar orbit. In less than an hour it will achieve rendezvous with the command module.

I am at the pilot's station, for Erwin was on watch almost all last night. A descent without incident; for an hour and a half the mesoscaph steadily approaches the bottom, making a few stops of its own accord. From time to time it hesitates before resuming its descent, even attempting to rise a little; the reason for this is that its inertia always takes it a little way beyond the point of equilibrium and, just as a pendulum swings back, it rises a little before assuming its position at the proper depth. This makes an amusing game of taking on water in the VBT at exactly the right moment in order to prevent this slight ascent without at the same time making the vessel descend too fast.

All this is very relaxed and goes on, moreover, to the accompaniment of popular music enthusiastically poured out by the cassette player. During the descent Don Kazimir orients the mesoscaph in a north-south position so that we can proceed as quickly as possible in the right direction in case there are currents on the bottom.

At 3:52 p.m. Ken Haigh announced that the bottom is visible through his porthole, about 10 meters below us. At 3:55 we are drifting rapidly along the bottom, crabwise for a few seconds, then, as soon as the guide rope has orientated us, straight forward. There is certainly a strong current; Don Kazimir's prayer has not been answered, but no doubt there are ahead of us enough fruitful observations to compensate us for the cold that we already feel spreading from the steel wall.

The water is not very limpid now; the bottom seems hard and granular. Would those be nodules of manganese, those strange agglomerations containing many kinds of minerals and appearing sometimes on the bottom of the sea like potatoes on a freshly harvested field? No, unhappily they are not. Closer observation reveals them as simply small, irregular undulations of the bottom itself. This bottom slopes gently; the depth continues to increase as we drift northward.

Since the water is fairly murky we pay special attention to the sonar which sweeps the area ahead of the mesoscaph. Its operation is clearly audible: a long whistle whose frequency alternately rises and falls and which reminds one of the early radio sets that one had trouble tuning to a station that was distant or too weak. We are all the more attentive because our chart carries a clear and suggestive inscription calculated to make us hope for some interesting encounter: "Careful, dangerous area." It is probably a location where the Navy, the Air Force, or the Army has at some time dumped old cases of ammunition or over-age explosives. If they are obsolete, probably they are no longer very dangerous. Nevertheless it would be better not to hit a percussion cap with our guide rope.

It commences to be cold, 15.5°C., but during the coming hours the temperature will go down another 5° or 6° at least, for outside it is now 8.36°C. This chill temperature apparently does not in the least disturb a charming little prawn swimming about 7 to 8 meters from the bottom. I cannot see if it has, like some crustaceans, swimming pads on its rear feet, making it possible to navigate freely in the water.

Observing the bottom more closely, I see that what I had taken at first for manganese nodules are in reality "ripple marks" on a very dark formation of sand, perhaps containing titanium, and at the bottom of the furrows thus formed there is lighter sand or sediment of more recent formation.

In some places this bottom, which is quite broken up, is traversed obliquely by wide bands as though moving objects had left the marks of their passage. The keels of submarines? Hardly likely this far down; we are close to 400 meters and only research submarines can operate at this depth.

At 6:20 p.m. we come to rest on the bottom. For several minutes everyone rushes from porthole to porthole trying to see something. I see two beautiful, long fish I cannot identify; side by side they advance, crabwise because of the fairly strong current that is carrying them. Perhaps they are eels, but I cannot say for sure. The ripple marks are irregular, definitely varied: in some places the bottom seems swept, cleansed, and polished by the current. In fact, we are now not heavy enough to resist the current; the friction on the bottom is more than compensated for by the effect of the water that drags at us; we too are proceeding crabwise, which is contrary to regulations—is, in fact, a cause for some concern. In the common interest we believe we hear Frank Busby say he has finished his measurements, and we blow out the VBTs to ascend a few meters and allow the guide rope to resume its proper function.

We drift once more a few meters from the bottom; then suddenly the sonar clearly indicates "something" straight in front of us. Since it is difficult to stop because of the current, we decide immediately to rise a little, and with two motors turned straight up the mesoscaph makes a leap of 30 meters to sail over the obstacle. Don Kazimir informs the surface. We are asked what the obstacle is. Don replies that he does not know, that we have not seen it—maybe it is rock, coral, almost anything—and he knows nothing about it.

In the report which the *Privateer* immediately sends by radio to the land, the word "coral" is transmitted more clearly than the others, and the news spreads. "At 400 meters depth the *Ben Franklin* has discovered a formidable coral reef." Then details are added about the form, the color, the nature of the coral, and the geologists and zoologists are amazed that such a coral reef has been found at this latitude, at this longitude, at this depth. This news will be repeated a hundred times. Perhaps someday we will have to organize a new expedition solely to find coral in this region, in order to stop making denials.

It is so cold (about 14°C.) that even Frank Busby, who is tough, is prompted by fatigue and the humidity to decide to call it a day and let the *Ben Franklin* ascend again. The evening is calm, the mesoscaph is drifting steadily north, slowly gaining altitude.

At 8:00 p.m. Chet announces, half in alarm, half in satisfaction, that he has found "bugs" in our cold water. This is the limit. This water, which has been disinfected and which contains so much iodine that it is practically undrinkable, is now contaminated. Kaz asks the surface what to do. The reply, which comes a little later, states that apparently these "bugs" (a word that covers bacteria, viruses, insects, and various other pests) are not dangerous; we can drink the water. Nevertheless we must beware.

As a matter of fact, this water is so bad that no one aboard will regret not drinking it. The hot water is good; it is spring water containing no disinfectant, for it was heated almost to the boiling point when it was introduced into the thermos tanks; we have enough for drinking purposes and to prepare our meals, for we have decided—contrary to what had been planned in advance—not to use it for washing. In short, cold water and bacilli for showers, hot, or rather, tepid, aseptic water for meals, tea, coffee.

After the slight vertical push (five seconds of air to drive a little water out of the VBTs) the mesoscaph has begun to rise, slowly at first (1 meter in about a minute) and practically without interruption, then faster, up to 3.50 meters per minute. This ascent presents us with a puzzle. Is it really possible that we are being carried upward in a stream of water at an average speed of more than 1.80 meters per minute? Or might a valve simply be leaking steadily and continuing to drive water out of the VBTs, consequently making us lighter and causing us to rise too fast? It is obvious that in rising the mesoscaph is growing warmer again, thus it is expanding and its absolute density is diminishing. However, this phenomenon is more than offset by the water, which also grows lighter as it grows warmer in the present temperature range. This is in fact the principle of our basic stability. The only tenable explanation, and the one accepted by Ken Haigh, is simply that we are caught in an important current that is moving toward the surface. To check this ascent, as we have done before, we will let a little water into the VBTs. Toward 9:15 this evening we are again more or less stabilized, but with a tendency to descend, since we have had to increase our weight beyond the point of equilibrium in order to check our ascent in the current.

Frank has not yet finished his observations on the bottom. By a sort of tacit consensus we had decided to shorten our stay on the bottom because the cold was practically paralyzing us, but the area Frank wished to study, a particularly interesting zone of geological transition, still extends below us, and we agree to descend to the bottom twice again, once around 4:00 a.m. and once the following afternoon. Thus, instead of remaining twenty-four hours in the cold, we will stay only a few hours each time and in between will rise and warm up in more clement waters.

This new excursion will take us to a deeper bottom than the last one (540 meters). While Ken Haigh and Frank Busby study the explosion program that the surface is providing for us, I can spend a long time looking through the porthole. It is black outside, without the floodlights, for it is now 4:00 a.m., and at that hour even in summer light hardly penetrates to this depth. Fifty meters before reaching the bottom we pass through a shoal of cuttlefish, those handsome cephalopods, about 20 centimeters in length. On several occasions one or another of them emits a sepia-colored cloud. Since we are drifting with the water, the cloud is visible for a long time and only changes shape very, very slowly. At first it vaguely resembles the cuttlefish itself: this has led to the belief that the cuttlefish produces it to trick its enemies, who take the shadow for the prey; in reality the more likely purpose of the cuttlefish—if it has one, of course—would be to blind its pursuer with this cloud of ink and to disappear more easily behind it. But all this takes place in darkness, at least at our present depth. Well then? This liquid, this sepia—does it have some effect on the fishes' sense of smell and does it thus trick the enemy by confusing the olfactory trail? Or even, like a cloud of aluminum particles thrown out by an airplane that wants to escape an inopportune radar beam, does the sepia in some fashion affect the radiations which are perhaps emitted by the cuttlefish, for example by modifying the electrical conductivity of the water? More prosaically, when the cloud begins to fray and break up, it reminds us of the Rorschach test that the psychiatrists wished us to take weeks ago in order to check our behavior during the thirty-day dive.

In conformity with the program, at about 6 a.m. on July 22 we ascend again to spend the forenoon in a less frigid zone, at about 350 meters depth, and at 1:15 p.m. we are ready to descend again.

It is now 11.5°C. aboard. Who called the Gulf Stream a warm current?

Toward 3 p.m. we are to descend once more; this will be, though we do not yet know it, our last trip to the bottom on this expedition. We were to have made three more, but later on the current was to carry us into too deep waters and our descent would leave us far from the bottom.

Here the bottom is sandy, or perhaps sedimentary; it is difficult from now on to distinguish precisely; the current is fairly rapid and once more we are advancing nicely toward the north. This last venture leaves us all with an unforgettable memory; each of us, I think, has had the good fortune to see something beautiful or interesting: a crab 25 centimeters wide; a small ray; an "enormous" fish—the word in this case can be understood to mean 30 to 50 centimeters.

This drift along the bottom with the guide rope continues thus for a half hour. It seems so easy, preceded as we are by our sonar advance guard. In general there are little "transverse" ripple marks running approximately east and west, intersected by wide and very shallow crevasses, running "longitudinally"—that is to say, north and south. The nature of the bottom and of its substratum is recorded by the instruments of Ken Haigh and Frank Busby who will study the records later.

Only Don Kazimir is unhappy; they are using too much of "his" electric current. Is it not up to me, the leader of the mission, to determine how the work and the resources are to be divided? What good is this dive if we make no observations? As a matter of fact, I know very well that Don, like the rest of us, is concerned only with the success of the mission and that he would willingly give us more current if he could.

At 4:00 p.m. of that same day we leave the bottom, to see it no more for the duration of this mission. On three occasions at intervals of three minutes we release compressed air for an average time of ten seconds into the VBTs. The result is again hard to interpret: the *Ben Franklin* rises, but systematically by stages, pausing briefly each time, sometimes even descending a few meters, then by itself beginning to rise again. To foresee precisely what will happen in the next minutes would require a computer that could deal instantaneously with all the phenomena involved, in particular the past, present, and future temperatures of the water, the hull and the interior of the mesoscaph; the densities past, present, and future of the water; the speeds also past and present; the state of the two VBTs and the exact weight of the whole mesoscaph; the factors of compressibility of the water and of the mesoscaph as a function of the ambient pressure; and of course the vertical component of the speed of the water. Only in this fashion could one predict the movements of the mesoscaph, which behaves like a pendulum oscillating around a varying point of equilibrium and subject to varying impulses, positive or negative, ricocheting from zones of different densities, retarded by its own inertia and bearing with it upward or downward the whole column of water in which it moves.

Toward 8:30 p.m. the *Ben Franklin* seems stabilized at 270 meters. It is still cold, 12.08°C. in the interior and 12° outside. This means that our hull is going to warm up a little and therefore we will rise at least a few meters more.

That night I went to bed early, and according to my logbook I slept very well.

Next day, Wednesday, July 23, I wake up to find the mesoscaph in a tumultuous sea—full of internal waves. Naturally we feel nothing at all and the most complete calm reigns around us; but a glance at the depth graphs recorded during these last hours shows large numbers of internal waves of a greater amplitude than we have encountered hitherto. After a rise of 30 meters in twelve minutes, without any action on our part, there comes a rapid descent of 50 meters in seven or eight minutes. But Ken Haigh tells us that in a submarine he once experienced internal waves of 50 meters in a period of two minutes. Besides, it is not very important (except as a constant source of anxiety), and the general direction we are following is good. The current is carrying us straight northeast, parallel to the coast of Georgia; our speed also is "good," a little more than 2 knots since last night.

I notice that a routine has been established on the telephone: at the beginning calls were always made according to a prescribed ritual that was written out and considered obligatory.

"*Privateer, Privateer,* this is the *Ben Franklin.* Over."

The *Privateer* would reply: "*Ben Franklin, Ben Franklin,* this is *Privateer.* Over."

Only then was one supposed to begin the real conversation. Now this has been radically changed, and it is not uncommon suddenly to hear a voice say on the telephone: "Hey Kaz, what are you doing down there?"

In general the acoustics are good. When we are too near the surface (when the ship has to be some distance away horizontally) or too near the bottom (when the echo gets mixed with the direct reception), communication is sometimes difficult; at moments the voice reaches us with the majesty of a bishop speaking from the pulpit of a cathedral, and at others it seems, on the contrary, to consist entirely of the most vulgar and unjustifiable epithets. We are not, however, the least offended. What amazes us (to be truthful, amazes me in particular) is the sparseness of news about the astronauts. To our specific questions, the answer is simply that all is going well and that they are due to return to earth on Thursday, something we already know, Chet May having brought with him a detailed schedule of the activities of Apollo 11.

Life goes on aboard, with internal waves succeeding and preceding periods of greater calm. On this day we suddenly rise on one occasion by 65 meters, only to descend again by 50 meters, the whole action taking place in less than an hour.

Frank Busby tells us that in the era when the U.S. Navy was under sail, an officer who came on watch had to wait half an hour before being allowed to alter the setting in any way; it was estimated that it took that much time to assimilate and understand fully the condition of wind, sea, and sail. We have tacitly instituted a similar rule: when one of us comes on watch he knows that if he blows air into the VBTs prematurely and without considering, for example, the mesoscaph's movements during the last minute, he runs the risk of lightening the vessel too much and of having immediately to perform the reverse maneuver.

Thursday, July 24. All goes well. But the night has been cold, the temperature was less than 14°C. aboard; the sea seems to be getting calm. We are again more stable. Moreover we are becoming accustomed to these internal waves and in some cases can foresee their amplitude and frequency.

This morning we check the atmosphere with particular care. The apparatus for detecting possible toxic gases is unquestionably, of all the machines aboard, the one that emits the most disagreeable smell. This is just one of the many paradoxes that surround us; another is that the device to measure the consumption of current, that is, to help us economize electric energy, itself consumes 10 percent of our total output. So its warnings must result in our saving more than 10 percent if it is to make any sense at all—and this is doubtful. Moreover, in a few days we will turn it off, having acquired by then sufficient understanding of the problem. We have already seen that it is the disinfected water that contains bacteria, whereas the undisinfected remains perfectly pure. There are other problems as well.

Between 12:50 and 2:03 p.m. we learn in a piecemeal fashion about the total success of the Apollo 11 mission—the splashdown of the command module, the inflation of the flotation collar, and the arrival of the three astronauts on board the aircraft carrier. Like everyone else, we are relieved, happy, and filled with admiration over this achievement.

43

ENCOUNTER WITH THE *LAPON*

The drift continues. At 32° 9′ north latitude and 77° 47′ west longitude, we are in the midst of one of the principal training zones for U.S. Navy submarines. A collision with one of them, which could be fatal to both of us, is not to be feared. The area we are crossing has been marked out in squares, each square with its own designation. Every submarine on maneuvers has to operate in its assigned zone; our own position is regularly reported to the Center of Naval Operations, and the latter provides an empty space around us of sufficient size to eliminate all risk. What if one day we forgot to communicate our position? For one thing, this is inconceivable, and for another I have good reason to think that the Navy follows our course with great care, independently of the information it gets from Grumman.

A few days later a nuclear submarine has been sighted in the distance and we are promptly informed.

At once we pick up the telephone and call: "Research sub *Ben Franklin* calling nuclear sub!"

The response comes at once:

"Nuclear sub *Lapon* here. *Ben Franklin,* how are you?"

For several minutes we converse. It gives us singular pleasure to be in contact with someone new, to hear a new voice on the telephone.

"*Lapon,* why don't you come down and join us? It's better here than on the surface."

The *Lapon* replies, showing that she knows very well who we are and what we are doing:

"*Ben Franklin,* we don't want to risk disturbing you in your research work. *Bon voyage.*"

44

TEN DAYS ADRIFT

On the evening of July 24 further explosions; the echoes are numerous and impressive; they last up to nine or ten seconds. This means that, before dying away, absorbed by the water and the depths, they have traversed about 15 kilometers. On receiving our bearings we notice that our speed has increased somewhat; it has averaged 2.3 knots during the afternoon. I try to make a brief résumé of the situation, the work accomplished during the ten days of our mission so far, and I dispatch a message to Walter Muench in Palm Beach and to our friend Jerry Kallman who is in charge of liaison in New York and Washington.

"After ten days navigation in the Gulf Stream, the situation is as follows:

"Everything aboard is going perfectly. We have

had no technical problems with the functioning of the *Ben Franklin.* Physical life, maintenance of order on the mesoscaph, exterior cleanliness of the portholes, all this gives complete satisfaction. Interior atmosphere, excellent. Morale of the crew, perfect. Everyone is in good health. Up to now we have made five excursions to the bottom; on three of them we drifted at a good speed, 10 meters from the bottom for several hours, with excellent visibility. Bottom interesting but very little life. Cuttlefish, crabs, prawns, and small fish. No big fish since the swordfish attack. During the free drift, very few fish, a few cuttlefish. Little plankton since the first nights when we saw beautiful chains of salpas up to 5 meters long. No 'Deep Scattering Layer.' The interior of the sea seems to be in continuous motion. Great underwater waves cause the mesoscaph to rise and descend as much as 50 meters in six minutes, demanding continual attention on the part of the crew. *So far, so good.*"

Friday, July 25. Since midnight the mesoscaph has been astonishingly stable at about 260 meters. How strange this sudden calm is. Nevertheless if one is to believe the speed we had last evening, we are continuing to drift normally toward the north or the northeast. This morning everyone is asleep except Ken Haigh and myself. The past few days have been quite hard and particularly cold. Kaz and Aebersold have worked all night and so have Chet and Frank, each in his own place. They are now taking a well-merited rest. Now we are at 262 meters. No problems.

45

EJECTION FROM THE GULF STREAM

No problems? At 9:20 a.m. the surface calls me: we have probably emerged from the Gulf Stream. Bill Rand asks me to wake up Kazimir. I haven't the courage to do so and I let him sleep a while longer. Bill then asks me to approach the surface. I blow air into the starboard VBT for five seconds; at 10:09, for five seconds more. The mesoscaph is reluctant to move; at 10:27 we are still at 258 meters. I blow once more for five seconds. The ascent begins very slowly. At 10:45 the situation becomes "critical": Bill Rand insists on talking to Kaz, who is the one he understands best on the telephone (certain voices do not carry well) and tells him that now the situation is clear. We are out of the Gulf Stream which at present is flowing some 28 kilometers to the east of our position.

The Gulf Stream has rejected us. On the twelfth day of our voyage. Immediately we discuss all the possible consequences and the probable causes. When we were preparing for the expedition, it will be recalled, two theses were presented: first, that there would always be a tendency for us to be thrown out of the Gulf Stream or at least to depart from the center of it at the rate of about 5 kilometers per day; second, that if we started in the center at a depth of about 200 meters in 15°C. water, we should remain there without difficulty for one month.

I myself think that these phenomena are not yet well enough known so that each detail can be foreseen. It is known that the current has numerous branches that rise and descend, turn and divide, rejoin and fuse by turns. But here also a computer would be needed to forecast anything with accuracy; most of all we would need someone to establish the program for the computer, and isn't that exactly what we are trying to do? For the moment it is imponderables that guide us; the slightest variation in attitude or altitude, in orientation in the current, can in an instant make us veer off to one side or the other. Who can say in advance where a single drop of water in a given river will go, or which will be the first drop to evaporate or to cling to the bulrushes on the shore? Being in the center of the current when we left, we had a better chance of remaining there for a longer time than if we had been at the edge of the Gulf Stream. Now after ten days we have emerged for the first time. This is an interesting fact in itself. We shall try to make our way back on our own, following directions given us by the *Privateer,* which along with the *Lynch* has made enough soundings in the ocean to have a fairly good general grasp of the situation.

At 11:23 a.m. Kaz starts up the two motors at slow speed, thus keeping our autonomy at a maximum. If we proceed faster we consume more current per mile. The electric meters emit a sinister crepitation: each ampere-hour produces an audible *tac* throughout the mesoscaph. One generally hears an occasional *tac,* but now they come in veritable machine-gun bursts. We do not have enough current at our disposal, however, for many corrections of this kind; the whole point is to find out if we can get back into the Gulf Stream, for our absolute speed at this moment is hardly greater than 1 knot.

Following the advice of the *Privateer,* we are now navigating at 100 meters depth, maintaining our depth with the propulsion motors. Erwin is a specialist in this maneuver. At 12:20 p.m., suddenly looking through the upper porthole, I *see* the surface and the waves. I glance at the manometer; yes, we are indeed at 100 meters. This visibility is remarkable, very rare, even though it is not continuous but perfectly clear only for a moment at a time. Obviously, the sun is playing with the waves. I have never before seen anything so clear in the water at this distance. The water temperature is 19.69°C. Inside the mesoscaph it is now 18°, but the temperature will rise quickly. Along with Ken Haigh, we have acquired the custom, if we wish to communicate the temperature of the water, of using a corresponding date in history. With the water now at 19.19°C., we say, "Treaty of Versailles."

Now we are at 90 meters and the water here is a half degree colder than it was at 100 meters a while ago when we were "in 1969."

Navigation so close to the surface presents a wholly new sight: in the first place, it is light; even without interior illumination in the mesoscaph I can read and write easily. Outside, too, the appearance of the specks of plankton, phytoplankton, and zooplankton is quite different; a procession of creatures a few millimeters in diameter shows a bluish luminescence of great beauty as they pass by; they actually sparkle in the water like diamonds in the sun. I have seen them before on other occasions but I do not know them intimately; fortunately these are "acquaintances" of Ken Haigh who identifies them as diatoms or radiolarians. Along with these are a multitude of salpas going by, and many other infinitely varied forms, jellyfish, copepods, pteropods, and siphonophores, which strangely resemble little twigs of green algae. A few big salpas too, and one or two pyrosomes, which at this time are again not phosphorescent. Technically the dive continues; we are proceeding deeper again at about 1 knot on a 125° heading—that is, in the direction the surface says will bring us back to the Gulf Stream. The temperature of the mesoscaph has risen: it is now 21.5°C. The sea is at 18.90° (the Eiffel Tower is one year old), but last night at midnight at a depth of 270 meters the water was 16.50°; at 4:00 a.m. we are at the same depth (272 meters to be exact) and the water is only 12.87°; this is clearly consistent with our departure from the Gulf Stream; outside it the sea is cooler.

At 3:30 p.m. the surface says we are close to the Gulf Stream and asks us to proceed in the same fashion for another hour. I, however, am highly skeptical, for I do not believe that our absolute speed can be more than 1 knot. We have covered at most only 5 nautical miles, and therefore we must be far from the 28-kilometer point we were heading for, unless the Gulf Stream this time has come to meet us.

Moreover, the surface is not altogether sure; the *Lynch* had to return to shore to replenish supplies and fuel; now it has to be given time to re-examine the situation, to make soundings with the expendable bathythermographs, and to determine the exact position of the principal current. So in any case we must wait for tomorrow, and during this time simply keep the *Ben Franklin* at correct depth.

At the beginning of the night we see one or two elongated fishes that we take to be eels. After all, we are not very far from the Sargasso Sea where adult eels go to die and from which emerge the leptocephalus larvas ready to roam the world.

During the night the situation is clarified bit by bit: the mesoscaph has simply been caught in one of the gigantic whirlpools that often appear on the edges of the Gulf Stream and drift off independently of the principal current to die by inches when they have lost their energy; only rarely are they taken in charge again by the Gulf Stream.

On the morning of July 26 the formal decision is made to ascend to the surface and have the *Ben Franklin* towed for over 50 kilometers. This is the necessary conclusion, from observations made by the *Lynch.*

46

NEW HYPOTHESES, NEW STUDIES

Naturally enough, this was the occasion for us to discuss the mysteries and caprices of the Gulf Stream. This current was "officially" discovered by Ponce de Leon in 1513 while he was cruising along the Florida coast between Cape Canaveral and the region of the Keys. Christopher Columbus described a current in the Bahamas which has not been identified exactly but which was perhaps a branch contributing later to the Gulf Stream. Benjamin Franklin was the first to study this current in a systematic fashion and to chart it, or to have it charted. Franklin's notion that the Gulf Stream was a river that crossed the Atlantic persisted for two centuries and in fact was only recently exploded. According to that theory, the current is produced by the trade winds blowing from east to west, which drive the waters of the equatorial Atlantic toward the Gulf of Mexico, even causing the level of the latter to rise and forcing it to flow out once more into the Atlantic between Cuba and Florida. Johannes Kepler, at the beginning of the seventeenth century, attempted to explain these currents and came close to discovering the tendency for any body on the earth to drift sideways from a meridional course because of the planet's rotation. This discovery was made by the French mathematician Gustave de Coriolis in 1835 and is named the "Coriolis effect." A year later, in 1836, François Arago explained the current by a difference in density between the hot waters of the equator and the cold waters of the North Pole, picturing a marvelous thermosyphon on a planetary scale. Matthew Maury, one of the masters of American oceanography, shared this view.

Alexander D. Bache, great-grandson of Benjamin Franklin, during the nineteenth century assumed the responsibility and initiative for a new analysis of the Gulf Stream, between the island of Nantucket (south of Cape Cod) and Florida.

Finally, since its beginning in 1930, the Woods Hole Oceanographic Institution has specialized in the study of the Gulf Stream, taking repeated measurements of temperature at thousands of localities. Guided by Columbus Iselin, director of Woods Hole, Fritz Fuglister, Worthington, W. Van Arx, Henry Stommel, and several other great names in American oceanography, a whole crew of scientists pushed along Franklin's trail, trying to interpret findings which were at that time considered definitively established.

Nevertheless it was the Woods Hole group that attacked the notion that the Gulf Stream was a river in the ocean. Acceptable for the region toward the south, off Florida, the idea no longer fits beyond Cape Hatteras where several branches seem to exist. Numerous observations were made of the eddies abandoned by the Gulf Stream on its own shores. Thanks to innumerable measurements, it has been possible to discern loops of the Gulf Stream extending farther and farther and growing rounder and rounder until suddenly they form giant whirlpools independent of the principal current. All this with lateral displacements of the Gulf Stream of close to 20 kilometers a day (observed especially in 1950) coincided exactly with the adventure which we experienced on July 25, 1969.

Everything seemed to contradict the idea of a "monolithic" current, and Fuglister considered that the observations made one think rather of a vast system of currents "which overlap like the tiles on a roof. . . . Each new current commences before the disappearance of the old one and is always born to the north of its predecessor."

This theory, though it still does not explain the origin of the "tiles" which are as though dropped on the heads of the classic oceanographers, gives a better explanation of the formation of the eddies thrown off by the Gulf Stream. It was because of these ideas that many scientists predicted that we would very quickly be ejected from the current. That we remained there for twelve days tended on the other hand to confirm the fact that, to the south of Hatteras at least, the current is relatively homogeneous.

It was at this time also that Woods Hole, due to the labors of Henry Stommel, deduced the existence of a current running in the opposite direction beneath the Gulf Stream. Pure theory, of course, but all the more remarkable when this "counter current" was later found more than 2,000 meters deep in the Atlantic. The discovery was due to the famous Swallow buoys which float free like our *Ben Franklin* and regularly emit *pings* enabling the surface (in this case the *Atlantis* of Woods Hole and the British vessel *Discovery II*) to establish their exact position and thus follow the current.

In the light of these researches and finds, our ejection today is definitely interesting. Yet how can one generalize from a single event in a milieu that is still little known? My only thought about it at the moment is that it is imponderables or unknowns that lead us on and can either retain us or reject us.

I was to find afterward, in my office at Palm Beach, a study of the Gulf Stream carried out by the U.S. Navy Oceanographic Office between March 30 and April 8, 1966. At the beginning of April that year the Gulf Stream, at a spot only a little to the north of the place where we had been expelled, developed a great loop toward the northwest which was obviously ready to separate from the principal current. There are many spots where the beginnings of similar loops can be seen on the maps produced by this same bureau and published monthly by the Navy.

How are these measurements carried out, and how is it possible to determine each month the course of the Gulf Stream when, not too long ago, it took months to follow just a small segment of the current?

This has become possible through a revolutionary new method that gives excellent results. Navoceano acquired in 1961 a super-Constellation named El Coyote. This aircraft had a long range (6,500 kilometers in twenty hours flying time) and was especially suited to missions of scientific reconnaissance. It has been equipped with an infrared hygrometer; a small meteorological station for measuring air temperature and especially air pressure; a special radar for measuring the height of waves; an infrared thermometer which measures directly and instantaneously the temperature of the surface of the ocean (at the depth of a dollar bill, we were told) in the range of −2° to +32°C., with an accuracy of ±0.4°C.; an expendable bathythermograph system which transmits by radio to the aircraft its depth reading (up to 300 meters) and the water temperature, between −2° and +35°C., with an accuracy of ±0.25°C. The infrared thermometer has been calibrated empirically, taking into account, of course, the column of air that exists between the aircraft and the ocean surface.

At the beginning of 1969 I had the good fortune to go along on a routine flight of El Coyote. The pilot has on his instrument panel a galvanometer connected to the ART (Airborne Radiation Thermometer). It is stupefying to the novice to see the needle of the galvanometer leap up suddenly by several divisions on the dial at the precise moment when the plane flies across the edge of the Gulf Stream, that is, where the thermometer detects a definite rise in temperature. At 150 meters above the sea the circle of measurement determined by the cone of infrared rays on the surface of the water has a diameter of 4.50 meters. In a single day the plane can thus secure data sufficient for the specialist to draw a whole chart of the Gulf Stream in a given area.

Even more, if an airplane can observe the evolution of the Gulf Stream, a satellite can do so too and can produce—because of its much greater height—a much broader picture. On May 15, 1966, for example, NASA placed the satellite Nimbus 2 in an almost perfect polar orbit at some 1,100 kilometers above the earth and making one revolution in 1 hour, 40 minutes, offering thus successively the entire surface of the earth to the eye of the satellite. Nimbus 2 was equipped with a relatively conventional television camera for operation during the day and with two "radiometers," one a "multichannel medium-resolution infrared radiometer" and the other a "high-resolution infrared radiometer." The first was for measuring the ratio of heat between the earth and the atmosphere, the distribution of water vapor in the atmosphere, and the temperature close to the earth; the second for determining the cloud state during the night as well as the surface temperature of the earth and the ocean. The latter sweeps the earth in a band 3,500 kilometers wide, which is enormous, and with a definition of 11 kilometers, which is still rather imprecise. The temperature is measured with a precision of $\pm 1°$ Kelvin, which is remarkable. The data are instantly transmitted to land and can even be expressed directly in numbers.

Thus it is clear that the Gulf Stream, besieged on all sides by satellites, by airplanes, by ships, and now by submarines, in the end will have to divulge everything about itself and in particular about its origin.

The satellite method also makes it possible to detect at a single glance large zones of phytoplankton, thanks to the chlorophyll which the infrared detector picks up, and thus ships are guided to where fish, equally aware of the presence of the plankton (though by different means), are about to gather.

47

SURFACING

On Saturday, July 26, at 9:27 a.m., Erwin has blown out the VBTs to bring us to the surface. Since the batteries have been put to considerable use, we realize that they would contain a good deal of gas and that therefore it is necessary to rise slowly to allow the gas time to escape. The final 50 meters should be made at an average of not more than 5 meters per minute. While the gas bubbles are forming and growing larger they expel water from the bottom of the oil bath, causing the mesoscaph to grow lighter. Then when the stream of bubbles leaves the mesoscaph some water is allowed in and the density is augmented once more, but on the other hand the weight diminishes according to the weight of the gas liberated. In practice one can distinctly hear the batteries gargling and some of us even believe that we can detect the effect on the trim of the mesoscaph when gas forms and disappears in the water. For my part I cannot guarantee the accuracy of this observation.

During this slow ascent Ken Haigh, Don Kazimir, and I are at the portholes. We see several magnificent sharks go by; I made no note of their dimensions, but I remember estimating that they were between 2 and 3 meters long, which is fairly impressive. They pass slowly—no one can tell whether they are peaceful or sinister—close to the mesoscaph, and we see them really well. One of them is probably *Ginglymostoma cirratum,* a nurse shark, a beautiful brown creature spotted with black. This species is frequently found in tropical waters, never exceeds 4 meters in length, and is not dangerous. A book we have aboard classifies sharks as follows: "Man-eating sharks; nondangerous sharks; dangerous sharks [not to be confused with the first category]." Nevertheless we warn the surface: "Attention, if you put divers into the water there are sharks present."

A moment later we see a barracuda. Busby is astonished. Isn't this a wahoo, similar at first glance, but usually found alone in the open sea, whereas the classic barracuda is thought to live in schools and closer to the shore? But no, its prominent lower jaw certainly identifies it as a *Sphyraena barracuda.* In spite of its legendary curiosity it passes close to us without pausing, almost without looking at us.

Then suddenly—we are definitely in an interesting zone—a superb hammerhead shark, one of those extraordinary creatures which have been endowed with eyes at the ends of incredible protuberances extending at either side of the head and thus apparently so contradictory to the laws of hydrodynamics. Despite this, what suppleness, what stunning grace, in the simple undulation of this monster whose ferocity and speed are notorious. It looks at us from one side (with eyes placed in that way, it has a right to), circles once or twice around us, and disappears. Hammerhead sharks eat everything—sextants have been found in their stomachs, even cans of food—but we are decidedly too big a mouthful, and the odor of oil from the batteries which it must smell does not tempt it nearer.

Apart from this, I see a small fish, no doubt a myctophid (lantern fish), which is known to circulate incessantly between medium depths, or possibly even greater depths, and the surface, and has perhaps accompanied us in our ascent. Suddenly it dashes at a speck of plankton but has to let it go, for the latter defends itself stoutly and remains extremely agitated after the attack. The salpas, witnesses of the scene, seem to gambol joyously, turning and circling about themselves, abandoning themselves frenetically to their looping acrobatics.

The two VBTs are empty, which is normal and a good augury. But the telephone is impossible. Kaz tries to ask if we can break the surface. A long monologue of gurglings is the reply.

Kaz counterattacks: "If you hear me, say nothing!"

Unheeding, the surface continues to gurgle, to the sea, to others, to the sharks, but certainly not to us.

At five minutes past noon the conning tower emerges on an absolutely calm sea. Radio contact is established at once. The first stage of the dive has been completed. Not without a certain bitterness, I must admit, we see the sun shining through the upper portholes and feel the rocking of the *Ben Franklin* which the *Privateer* at once takes in tow. We move toward the east, toward the Gulf Stream which at the moment lies 50 kilometers away from us.

Of course we will not open the hatch. It is essential that the survival operation should continue rigorously intact, and even to the point (which seems to us greatly exaggerated) of our not being allowed to use the airlock that could transmit personal letters to our families.

By using the lock correctly, always keeping one of its doors closed, we would not have changed interior conditions any more than by using it at a depth. But, for the surface, orders are orders, in spite of all clear evidence that they are pointless. This can be understood: communication with the land is bad, often very difficult; they prefer to keep to fixed general rules rather than admit improvisations which in the present case would have done no harm but on a different occasion might chance to be dangerous.

So we remain during that day, in fact for exactly sixteen hours, on the surface, which we have done everything possible to avoid for one month. Through the portholes we see from time to time a member of the *Privateer*'s crew walking on the bridge of the mesoscaph, carrying out certain checks, removing the proton magnetometer that ceased functioning after the first few days of the dive. This is an expensive apparatus and Busby is anxious to have it safely stowed away as soon as possible.

During this long day and the early part of the night there is nothing in particular to do aboard. Each of us is thoughtful, wondering what the future holds. Since we were ejected from the Gulf Stream after twelve days, even before reaching Hatteras, what will it be like beyond? Will we succeed in remaining in the current for twenty-four hours longer, or forty-eight hours? What useful work can we still carry on during the second part of our mission, which seems to be in danger of becoming a series of separate, small missions?

In fact, putting together different pieces of information received bit by bit from the surface, we begin to understand how badly we have missed the *Lynch* in recent days. With its brilliant crew of oceanographers aboard, it would probably have been able, had it remained in position, to detect the unexpected movements of the Gulf Stream and, had we been informed in time, we might have avoided ejection from the principal current by getting under way ourselves at the beginning of the phenomenon. This is the important lesson of the event. If only the *Lynch* can now accompany us to the end.

As I have said, we do not expect to open the hatches. There are several reasons for this. In our discussions on shore before departure, it had been agreed that even if the drift experiment had to be modified by circumstances beyond our control, the survival experiment should continue if possible up to the limit of thirty days. NASA especially, and consequently Grumman, which had a formal contract with that organization, was more concerned with these thirty days of survival than with the 1,500 nautical miles to be traversed under water.

Also I think we would all have had trouble taking up the "burden" of isolating ourselves again in the mesoscaph for two or three weeks if we had tasted a semblance of liberty that day in the sun and fresh air.

Dr. Alain Bombard, who crossed the Atlantic in a small lifeboat and sustained himself entirely on what he could catch in the sea with whatever means at hand (well studied in advance but simple enough so that theoretically any shipwrecked person can make use of them) has related that at mid-voyage he encountered a liner and was invited aboard. Thinking he could interrupt the monotony of his voyage without disadvantage, he accepted. Naturally he touched no food while aboard the liner, happy just to be in human company once more. However, upon returning to his rubber boat, he had difficulty getting used to it again, and the interruption, instead of being salutary, was a disagreeable sort of illusion which took a long time to dispel. With us the problem is somewhat similar: I prefer, and I believe we all do, to remain shut up in our underwater abode without interruption for the whole thirty days.

In any case, we can absorb some heat. With the sun shining on the hull, the interior temperature goes up to 29°C.

One further small incident remains to be mentioned.

48

FREE FALL

We had come to the surface by using the motors and regulators and theoretically we could descend again by the same means. But it was advantageous to take aboard ballast principally to make up for the weight of the air we had used, which amounted to almost 150 kilograms. To avoid using our motors and squandering electric energy, I requested the surface to add 450 kilograms of ballast to give us an excess of 300 kilograms, which could easily be dumped during the descent. The surface replied that they would give us 1,125 kilograms, just to be on the safe side. Afterward they explained that they had wanted to be sure the descent would begin easily and that there would be no possible need of adding ballast at the last moment (which, to be sure, at night and with possibly a heavy sea, would have been troublesome).

As everyone knows, there can be different ways of attaining the same goal. In this case the surface took the initiative. Their aim was to be sure that the descent would begin without fail.

It began well but fast!

4:07 a.m., July 27, beginning of the dive.
4:11 a.m., 110 meters.
4:15 a.m., 250 meters.
4:19 a.m., 400 meters.

From the beginning of the descent we have been dumping our excess iron shot, 725 kilograms since we left the surface, and we are still descending rapidly, though slowing up a little. Given our original speed—one felt as though in free fall—the mesoscaph will descend deeper than its point of equilibrium and rise again, then will oscillate for some time before stopping at its depth of stability. We must not try to stop it all at once but to estimate its speed and its weight, the relation to the temperatures, problems that we are working on now, not always fast enough, but at least mastering the situation reasonably well. Nevertheless Kaz is worried, not because of any danger to the mesoscaph, but for fear that a false maneuver might cause us to rise to the surface again or make us exceed the depth to which "we are entitled."

At 425 meters the *Ben Franklin* stops, hesitates an instant, and, as foreseen, rises again, to 346 meters, then all by itself begins to descend. Now all is going well. This is the pendulum that will come to rest. Nevertheless by slow degrees we dump another 175 kilograms of ballast, which makes a total of 900 kilograms. We had asked for 450 kilograms with the intention of dumping 300. They gave us 1,125—675 more than necessary—and we have dumped altogether 900 kilograms, that is, 600 more than we would have needed. We have come out right to within 75 kilograms. These 75 kilograms, however, puzzle me. But after the dive I learned that at the last moment the surface relented and reduced the added load to 1,050 kilograms which explains the discrepancy, and so our calculations were not so far wrong.

After more oscillations and a gentle re-ascent by stages, the *Ben Franklin* proceeds to reach perfect equilibrium at between 200 and 210 meters depth, in full accord with its basic stability. The day has been a full one. Those free from watch duty enjoy a well-merited night's sleep.

July 28, another deep descent. There can be no question here of coming to rest on the bottom, for the sea is too deep (1,400 meters as we begin our descent). Frank and I hold a long discussion about the timeliness of this descent. On the one hand we know that by leaving the "branch" in which we have been so well settled since yesterday, we will obviously find a different stream when we come up, for the current will be less rapid between 500 and 550 meters than it is here at 200 meters. During our descent, if the part of the Gulf Stream where we are at this moment displaces itself laterally, we will lose it and so will risk having to surface once more. On the other hand, the Gulf Stream is more than 110 kilometers wide here and we are well placed at 35 kilometers east of its northwest "wall." Frank estimates that there is little risk of losing the current in the course of a dive of ten to twelve hours. We agree that the risk, which, though small, unquestionably exists, is well worth running if we take into account not only the precise measurements that we are going to make but also the theoretical interest in knowing whether we will be ejected once more.

49

DEEP-WATER DIVE

A little after noon I reopen the exhaust valves of one of the VBTs and the descent starts, intentionally slow, to give us a chance to make visual observations through the portholes. At 3:00 p.m. we are at 500 meters; at 4:20 in good equilibrium at 570 meters. From then on the *Ben Franklin,* continuing to cool, has of course a tendency to keep descending; also it is necessary to watch the depth with close attention and at intervals to blow a few seconds of air into the VBTs to compensate for the increase in density—some seventeen times in all over a three-hour period. During the entire time Frank and Ken are recording the explosions from the surface, the mesoscaph varies barely $\pm$ 1.50 meters. In fact, this descent into deep water makes a particular impression on me, one that military submariners would not understand. For the first time in my life I am in a submarine which *must not* descend beyond a limit this side of the bottom. Up to now, with the *Ben Franklin,* with the *Auguste Piccard* in Lake Geneva, and with the *Trieste* in the ocean, the bottom always represented a safe place where the submarine could come to rest, the hull being strong enough to withstand any foreseeable pressure. Now, as with all conventional and atomic submarines operating in the high seas, we must be careful not to go down beyond the limit of safety, for the hull would implode without warning.

Kaz is talking on the telephone. He ends with the word: "Okay?"

The reply comes immediately: "Okay!" But it comes so quickly that we realize it is an echo from the sea and not the approval requested from the surface. In this instance it is unimportant, but there are many cases when echoes could create serious confusion—all the more so since the strongest and clearest echo is not always the first to reach us.

A more worrisome subject is the relentless battle against bacteria; now we have to put disinfectant almost everywhere: on the floor, on the sides of the shower, in the galley sink, even on the walls of the mesoscaph. Often we gain ground; often we lose it. It is truly a race against time; we must hold out fifteen more days. Will we manage it? Chet organizes the struggle, supported by Kazimir, and everyone else collaborates in one way or another. For the moment the hot water is still good and aseptic.

During the night we reascend and I note with relief that in the neighborhood of 400 meters the outside temperature is identical with what it was during the descent. We hope that this will continue. If the temperature is the same, that means that we are encountering practically "the same stream." Next morning, however (July 29), when we learn our position from the surface, we see that the dive has changed our direction; in the depths we have drifted straight eastward some 7 kilometers, whereas at 200 meters our direction was substantially northeast. The next days will show us, however, that we have been safely inside the principal current throughout.

There are no internal waves, and for six hours we remain at exactly 200 meters with no oscillation exceeding 1 or 2 meters at most.

50

AT HALF TIME

This evening we pass the halfway point of our mission. We have been slower than anticipated, but this has not affected the scientific work accomplished. We emerged from the Gulf Stream once by missing the curve of the Carolinas, but we got back safely and since then the mesoscaph has maintained a good position. Altogether the *Ben Franklin* has behaved remarkably well; the batteries, those elements immersed in sea water, are doing well too, and now there is every likelihood that they will produce the full capacity—750 kilowatt-hours—expected of them. The morale too is good; differences of character have not resulted in any serious disagreements. Indeed, up to now the operation is satisfactory.

Nevertheless I regret the absence on board of the complete calm that I expected and that, it seems to me, would have permitted even better working conditions. Without striving for the devotional silence of a monastery, we might still have had some hours of quiet concentration so fruitful for scientific research. But concentration would be possible only if the cassette player broke down. It is a good instrument, however; at first the others were afraid its batteries would run down too soon, but Ken found a stock of reserve batteries that would serve the purpose. There is nothing to be hoped for in that quarter. So much the better for those who love modern jazz and western music.

On several occasions I have had good and peaceful hours; often in the morning everyone is asleep except Ken and me. Ken is silent by nature and entirely dedicated to his work. When I am alone with him on watch everything is commendably quiet and tranquil. No sound disturbs the atmosphere except the "Gertrude" check every half hour inquiring about our temperature and depth. For a while the surface really needed this information. Now I suspect it is just routine.

By regulations I should awaken Don or Erwin at midday to take over the watch. At first I did so out of respect for the established order. Later on I came to appreciate this beneficent and useful period of calm so much that I postponed wakening them and finally abstained completely from having them relieve me. There were advantages on both sides: Don and Erwin, who worked a great deal at night, gained some hours of well-earned rest and I myself could carry on my work in much better circumstances.

On the surface the weather is bad. The sea has risen to Sea State 3 on the Beaufort scale. The wind is blowing at 10 to 15 knots; according to the scale this is a "moderate breeze," or a "jolly breeze," but the surface doesn't find it jolly. It is raining too; we can hear it clearly on the acoustic telephone. We picture the *Privateer,* our "pirate" escort, rolling and pitching, all the worse because being on the surface it spends half its time drifting toward the north or northeast a little faster than we do and the rest of the time has to proceed south or southwest to rejoin us, whatever the direction of the waves and wind. It has to take the sea as it comes, wherever it has to go, since this is the only way it can constantly follow us. The *Privateer,* packed though it is with scientific apparatus, is nonetheless a mere cockleshell of poor stability, a light wooden vessel made to pick up mines, with a crew of enlisted men who were ordered simply to go out and to come back someday if possible. But this crew for the past fifteen days has consisted of volunteers, aware that within a month the sea can offer every sort of surprise, yet ready to accept whatever comes. Some of these men are accustomed to the ocean; others much less so; but all are there to aid us, to make this dive possible, to help us accomplish the Gulf Stream Drift Mission. If credit is due for endurance, it belongs, not to us in the calm depths, but to those who rolled and pitched at every whim of wind or sea.

Don, upon awakening, brings out a copy of *Life.* Every five days he shows up with a different magazine; by distributing them at calculated intervals, he attempts to give us a sense of reality. This time we decide to turn the tables.

"Did you notice in the last issue of *Life* the extraordinary pictures that Armstrong and Aldrin took on the moon?"

"No," says Don, caught off guard, and he searches feverishly for several seconds without realizing that this issue is three weeks old, printed well before the departure of Apollo 11.

What comfort, what security, to be able to work in such an atmosphere.

The mood aboard is good. I begin reading an article by Ken Haigh, given to me earlier so that I could appraise the seriousness of this scientist and decide whether he would be the right person to accompany us. Having met Ken, I had no need to read his article to form an opinion; in fact, I had no time to do so before our departure. In his article, speaking of the harassing problem of scientific equipment's never being as modern or as efficient as he could wish, Ken Haigh set down this piece of wisdom: "A true scientist or engineer is never completely satisfied with his instruments." That evening, talking about the efficiency of the mesoscaph, I casually ask Ken whether he is satisfied with his equipment here. He replies, "Yes, of course," and looks at me in astonishment when I ask him if he does not consider himself a "true scientist." Then I put his own article in front of him. Good humor has reigned almost without interruption during the entire expedition.

Suddenly we hear, dully at first through the telephone, then more and more clearly and directly, a strong, harsh rumbling all around us: it is a huge freighter passing quite close to us, perhaps directly above us. We all fall silent with the same instinct—anguish apart—that caused submariners to fall silent when, during the war, an enemy destroyer stalked close and might from one instant to the next drop its depth charges. We have our own explosions too, but they are not 500-kilogram mines; they are simply either the "caps" of less than 30 grams fired electrically by the *Privateer* just below the surface of the water, or "suss" (submerged underwater sound sources) fired by the *Lynch* with a sort of 60-millimeter mortar and weighing about 450 grams each. These explosions, incidentally, are a new form of solar energy, reaching us from the surfaces; exploding, the powder gives back the energy absorbed during its composition. The detonation of the suss is so powerful even at a distance of some kilometers that one gains some conception of what a real depth charge in immediate proximity to a submarine would be like. Often the submarine is destroyed not by direct impact but by the rolling that results from the explosion, rolling of such violence that the acid of the batteries is spilled and the submarine is helpless for lack of electric power. On one occasion one of these suss fired more than a kilometer away caused an object to fall off one of our shelves. After that I requested that they increase their distance slightly for future explosions. I recall, moreover, that although our steel is similar to that used in the hulls of most American naval submarines, it is not precisely the same—that is, it is less resistant to shock.

July 30. It is now up to Frank Busby to send a report to Washington, informing his chiefs of the work accomplished up to now by Ken Haigh and himself (apart from that of the rest of us):

More than two million measurements of salinity, of temperature, and of sound velocity as a function of depth and time of day.

6 hours of measuring the current on the bottom

848 pairs of stereoscopic photographs on the bottom

45 hours of measuring the ambient light

14 hours of gravimetry

5.5 kilometers of recordings of the Side Scanning Sonar

425 experiments in acoustic reflection on the bottom

90 hours of direct observation

All this in the course of 360 hours and 1,000 kilometers of drift.

Through the portholes quantities of plankton are again visible: *Phronima, Pyrosoma* with numerous reflection points and trailing innumerable tentacles, a few shrimp with very long antennas, several cuttlefish that come diving straight at us, Stuka fashion, discharge a cloud of ink, and disappear into the night. One of them, however, comes to rest on the frame of my porthole, this frame being a plastic cone with an almost horizontal lower edge. Although practically independent of gravity, the cuttlefish rests there a moment, and for this interval all of us can observe it under excellent conditions. Its eight arms and two longer tentacles are turned toward the porthole; its two "stabilizers" are on the opposite side. When it leaves, I observe a small creature literally enveloped in a cloud of ink which clings to its body and keeps it from moving. Even if it were free, where would it go? Here or elsewhere, are they not the same thing down where we are? Could this cloud from the cuttlefish serve to paralyze its intended prey? Perhaps that is its function; I must talk to a zoologist about this when we return to the surface. But what an unquestionably marvelous system for observation is this apparently simple combination of portholes and floodlights, which in fact made possible the creation of the first bathyscaph. If Plexiglas had not been invented (in 1936 by the firm of Röhm and Haas in Germany), what could have served instead? Ordinary glass, tempered glass, quartz are all too hard and brittle for big windows. A satisfactory result might have been obtained with quite small portholes and a system of optical fibers to transmit images at a distance. However, as soon as one makes use of lenses or any other complex arrangement, much of the advantage of direct visibility is lost. Submarine television, which could theoretically make it possible to dispense with portholes, never gives the same effect as good vision, never produces images of high definition. Only real portholes dispel the feeling of imprisonment and give the sense of really being in the element to be observed.

July 31. The event of the day is Don's decision to rewarm the water in Tank No. 3. Finally we have tea that is really hot. Incidentally, we have been making 2.14 knots during the night. It is a record.

Tonight I once more look out the portholes for a long while: in the dark I see a series of luminous points, natural bioluminescence; several times I turn on the light suddenly to try to surprise the animal that carries these lanterns, but I see nothing. They must be totally transparent. There are a few shrimp *(Sergestes),* most of them 3 to 4 centimeters long, a few at least 10 centimeters; a number of *Sagitta,* some euphausids, and big salpas. These last trail long tentacles that seem to be retractable; only a part of their bodies, the rear half, carries these tentacles; this half contains organs that are clearly visible and strongly reflective; the rest of the body is completely transparent. I try the plankton tube once more but do not succeed in catching anything of note.

It is not quite so cold, or perhaps we are accustoming ourselves progressively to this temperature, which certainly is much better here (20°C.) than in our overheated offices in wintertime or our superchilled air-conditioned quarters in Florida. For more than twenty-four hours we have been absolutely stable, between 160 and 170 meters in depth.

The surface calls us. It is Paul Campbell's clear, distinct voice, which both we and the sea prefer because it travels without interruption along the carrier wave of our acoustic telephone.

Ken Haigh picks up the instrument: "*Ben Franklin* here. Over."

"Your temperature and depth?" Paul asks.

"Temperature?" Ken asks me, for he is in the bow and I am in the stern where the thermometer is—or more exactly, its indicator.

"Accession of Louis Philippe," I say.

"Eighteen point thirty degrees centigrade," Ken translates for the surface.

"I read you loud and clear," Paul says, and we go back to work.

During the evening we rounded Cape Hatteras, passing 70 kilometers from the coast. Since our adventure off in the Carolinas we have been following the current with no difficulty. Cape Hatteras certainly marks an important stage on the journey, but what now lies in store for us?

51

NATIONAL HOLIDAY

Today, August 1, is the Swiss national holiday. The two Swiss aboard, Erwin Aebersold and myself, must celebrate; to be specific, we *must* have a fire; it is a tradition, almost an obligation. The August 1 fire, which was revived in Switzerland in 1891, the six hundredth anniversary of the Helvetic Confederation, is a reminder of the fires that served, in place of a clandestine radio, to arouse the people against their foreign oppressors. This happened at the beginning of August in the year of grace 1291. The rebellion was successful; therefore it is permissible to celebrate it. If it had failed, no doubt we would have fires to commemorate the defeat of the anarchists who in their blind irresponsibility had risked the parceling up of their native land. In history as everywhere else it is only success that counts.

How can we carry out our rite? The bare idea of having a fire would make Kazimir tremble, and the surface is certain to forbid it. Nevertheless we have a fire. Tradition does not say that it has to be a big one out in the open air; it simply says that there must be one. Since tradition is a form of order, Kazimir, much to my delight, and I must say to my great surprise, offers no serious objection. So we solemnly display the Swiss, American, and British flags, then Erwin and I strike a match. The little flame dances, astonished at being surrounded by water, astonished by the attention it attracts; it dances for some seconds, then subsides and goes out. We have had our fire. In the annals of tradition this is probably the first ceremonial fire 200 meters down in the middle of the Gulf Stream. Kaz simply informs the surface that in observance of the Swiss national holiday, "Jacques and Erwin carried out pyrotechnical activities." This seems so improbable that the surface considers it a joke. We also send a message to the President of the Confederation and are deeply pleased to receive a warm response.

How much carbon monoxide was produced by our match? An infinitesimal quantity, but this problem of carbon monoxide is beginning to become acute nevertheless. On July 24 we had ten parts per million in our atmosphere (about 1 gram altogether). Last night we had twice that amount; if this increase continues we run the risk of exceeding the authorized limit. Carbon monoxide, unlike carbon dioxide produced normally by respiration, does not have the advantage of announcing its presence through preliminary headaches. It combines with the hemoglobin in the blood, which has a greater affinity for it than for oxygen. Since hemoglobin normally transports oxygen throughout the body, when it is thus overloaded the blood becomes incapable of performing one of its essential functions. Little by little the body succumbs, the victim often unaware of the inception of this process. Thus it is essential to measure regularly the proportion of carbon monoxide in our atmosphere.

Whence comes this carbon monoxide? Doctors have told me that it is normally produced in extremely small quantities along with carbon dioxide during respiration, but others have maintained that human production is so small as to be negligible in our case. Apparently, then, we have aboard some other source. This might be, according to other experts, a spontaneous emission from certain plastics, the electrical insulation, for example. In any case, we try out a "decontaminator" we have with us: an apparatus that "burns" carbon monoxide, turning it into carbon dioxide which will in turn be absorbed by the trays of lithium hydroxide. A precise measurement in our atmosphere before and after shows that the apparatus absorbed no carbon monoxide at all. (Later we learn that it would have functioned only in a dry atmosphere, probably in a relative humidity of 50 percent at most, whereas at that time we had about 70 percent.)

This evening, to take the place of the choruses in national costume that sing in the public square of the cities and hamlets of Switzerland, the sea offers us a generous and remarkable concert of porpoises. Unhappily we do not see the performers. I believe too that I recognize the barking of whales, which is not unlikely, since the surface tells us later that one was sighted at this time.

52

COMMUNICATIONS

This whale will be the cause of a lot of talk. The *Privateer* announces to shore control that they have seen a whale. Shore control does not understand. *Privateer* repeats the message. The word "whale" does not get through; it is understood as "wave."

At wits' end *Privateer* says: "No, no, a whale—you know, those big black fish!"

That gets through. It is announced in the press that the crew of the mesoscaph has seen a huge black fish in the Gulf Stream, and once more we will have to deny the report a hundred times. It's strange; one is able to converse with men on the moon, where they have only very small and ultralight transmitters. Between the *Privateer,* which has no weight limit, and the land, which can have any kind of receiver, the radio continually plays tricks of this sort.

The word "fish," which is confusing here, being of course wrong when applied to a whale, is nevertheless forgivable. Herman Melville, author of *Moby Dick* and an authority on whales, employed it often himself. Moreover, several species of small whales are called blackfish, even though they are mammals.

On this day too (August 1) we release for the surface one of our Plexiglas spheres containing certain messages and some samples of bacteria intended for Dr. Jessup, a member of the surface crew. From below we hear the *Privateer* maneuvering to pick up the sphere with a long-handled landing net brought along for this purpose; the sphere itself is easily seen, thanks to its intermittent flash, which Dimitri Rebikoff of Melbourne, Florida, arranged for us; thanks to this flash, we too, with our own eyes, are able to follow the sphere through the upper porthole during a large part of its ascent. As far as we can follow it, it rises vertically above us, showing that there is no appreciable difference in the speed of the current between 160 meters depth and probably about 100 meters. For the moment, too, we are continuing with what we call "good speed," a little more than 2 knots, which is more than our average before our "expulsion" off the Carolinas. What is more important, the direction seems good; at most, a little more easterly than the average seasonal position of the heart of the Gulf Stream during recent years.

Basically all goes well; we have no difficulties; after a few days, our emergence from the current last week seems a commonplace incident. Our survival poses no obvious difficulty (apart, perhaps, from bacteria and carbon monoxide). Is it surprising that we are in such good health, able to continue our work without difficulty, to pursue our course at the behest of the current? On the contrary, isn't it a triumph of modern technique to be able to traverse 2,500 kilometers under water in thirty days and without trouble?

August 2. During the afternoon the *Lynch* again determines our position in respect to the Gulf Stream. It is excellent: 20 kilometers to the northwest of the center of the current and 20 to the southeast of its northwest "wall," which is generally called the north wall.

53

ANNA THREATENS

At 7 p.m. today a surprise and a full-scale one at that: recording our absolute position on the chart and comparing it with our last position, we discover a speed of 5 knots. Even allowing for a certain impreciseness in the reading of the Loran apparatus, we would be making at least 4 knots, in contrast to the bare 2 knots that was our speed on the preceding days. We are skeptical, but we have to realize that Woods Hole has *calculated,* taking into account the differences in temperature, pressure, and other factors, that in this region speeds of 5 knots and even 6 are possible. Up to now no one to our knowledge has encountered such speeds. Whatever it is, we will see whether it keeps up, and of course each of us waits with impatience for the new bearings from the surface.

Our speed may become a factor of primary importance, for today we were told of the formation of a tropical storm, a hurricane, the first of the season. It has been christened Anna, according to the system by which these storms are given feminine names in alphabetical order, the first of the season commencing with A. At the moment Anna is 150 kilometers southwest of us, almost exactly where we rounded Cape Hatteras. Does the menacing title "hurricane" actually apply to this disturbance, which anywhere else would probably be described as a storm? The sea off Hatteras is famous for its tempests. Jules Verne, in *Twenty Thousand Leagues Under the Sea,* referred to this region as "the homeland of waterspouts and cyclones which are engendered by the current of the Gulf Stream."

Everything now depends on knowing what Anna is going to do. Theoretically her most probable direction is exactly the one that we ourselves are following; supplied with energy by the hot water of the Gulf Stream, Anna may be able to overtake us, increase her power, and become a serious threat to the *Privateer.* What then? If the *Privateer* has to leave in order to seek shelter, we will have to ascend and leave with it, because the Navy and Grumman would not want us to be alone in such a situation. With the *Ben Franklin* in tow the *Privateer* would be almost paralyzed in a storm, and so we should leave *before* the tempest strikes us. But as no one knows *when* the storm will come, or even *if* it will come this way, it is difficult to decide when the *Privateer*—along with us—should leave this area. Actually, in view of the suddenness with which Anna and her sisters move, the safest thing to do would be to flee as soon as a hurricane appeared in the region. This, however, does not help us with the problem, for no one knows what the "region" is. Before departure the simple provision was laid down that we should surface if a hurricane *arrived.* Has Anna arrived? No one knows. She has been born, she is alive, she is menacing, and somehow for want of the ability to reach a logical decision we go on.

Right now Anna's menace is not deadly, however disagreeable it may be.

The next day, Sunday, we recalculate our speed: yesterday between 7 a.m. and 7 p.m. our average was 2.5 knots. During the night it increased to 2.85 knots, which is less than the average for late afternoon yesterday but considerably above our general average. No more news of Anna. So much the better, especially for the *Privateer;* apparently the wind on the surface has not exceeded 40 knots, which is already a strong blow. The state of the sea rose to number 4, which does not agree with the Beaufort scale wherein 40 knots ("fresh gale") corresponds to a surface state of number 8. As is well known, estimating the state of the sea is difficult. In this case I believe 40 knots was reported by radio and if the state of the sea locally was really number 4, it meant a local wind velocity of some 16 knots at most.

What an idea—staying on the surface! How privileged we are to spend this month under water, and what a demonstration of the advantages of the submarine over oceanographic surface vessels. Will the time ever come when all cargo ships, transatlantic liners, tankers will be submarines, with all the implied advantages of regularity of service, of comfort, speed and, most of all, of safety?

In regard to the surface, we have just learned that Thor Heyerdahl, wishing to demonstrate the possibility of a migration from equatorial Africa to South America by allowing himself to be carried on a raft by the north equatorial current, has almost arrived at his goal. The *Ra* is constructed exclusively of materials that were available in Egypt three thousand years ago. Although he was finally swamped by a storm, he had time, nevertheless, to show that with luck the ancient navigators could very well have sailed from Africa to America. I would very much like to cross the Atlantic someday, perhaps with Heyerdahl, but only in a submarine.

The explosions from the *Lynch* produce fantastic acoustic effects. We now have 4,000 meters of water under us, and the echo, probably rebounding from different underwater valleys over which we float, lasts for about twenty seconds, the time required for sound to traverse more than 30 kilometers. This afternoon muffled explosions could be heard at a distance, coming perhaps from another group of researchers, or from workmen tens or hundreds of kilometers away.

When sound waves are emitted in what are called sound channels or acoustic channels, from which the differences in density prevent them from escaping, they can travel thousands of kilometers and have sometimes been picked up from one end of an ocean to the other. It has been noted, for example, that a kilogram of explosives detonated in the Hawaiian Islands can be detected as far away as California. If we are in one of these channels, the small explosions we hear may just as well originate on the coasts of Africa or Europe as those of the United States.

54

THE PRISONER SALPA

I spend part of Sunday afternoon with Don Kazimir, taking apart the pump for the plankton tube. I do not know whether it works better now but in any case it works, something I had not been sure of before. This evening at 10 o'clock, after keeping lighted for three quarters of an hour a 250-watt bulb placed on the outside close to the opening of the tube, I go to work again under the ironic and ill-dissimulated smiles of several of my comrades. With one hand I operate the pump, with the other I direct a strong flashlight on one of the small portholes in the apparatus, and through another porthole I look on, waiting for something to happen. This time I plainly see that the water is moving rapidly in the tube, and through the large porthole in the mesoscaph I also see that life is teeming outside. It is impossible under these conditions not to capture *something.* Pretty soon I see a salpa passing through; instantly I stop the pump and close the two valves, upstream and downstream, keeping the salpa prisoner in the central chamber. Well now! It is my turn to exult: I summon all those who are not asleep or at work to see the capture. What a pleasure, I would even say what excitement, to have one of our little friends so near without doing him the slightest harm. He barely seems to know that he is a prisoner. He swims in the chamber, moves from one porthole to the other, breathes, inflates himself, deflates himself, gambols as he was doing in the ocean, now a few centimeters away. The demonstration is conclusive; the system works. Even Frank is convinced. I send a message to Dr. Fye at Woods Hole to commemorate that scientific meeting in Florida at which he suggested the construction of this apparatus.

Next day (August 4) an unexpected visit by telephone: Don Terrana, the first Grumman engineer who came to help us in Switzerland with his valuable advice and who followed the whole construction with infinite care and perseverance, has arrived aboard the *Privateer* by way of the *Lynch,* on which he embarked while the latter was in port for supplies. His visit gives me special pleasure; he is the Grumman engineer who knows the *Ben Franklin* best. We can usefully discuss a number of technical problems on the underwater telephone.

Our actual speed and the regularity with which we follow the Gulf Stream astonish everyone. We are now making 3.2 knots. Never before have we been so far from the coast and we are heading straight toward Nova Scotia, even approaching once more the "average course" of the Gulf Stream. We are so far from land that we no longer know what our port of debarkation will be. Frank Busby, who is from Washington, hopes it will be Norfolk at the entrance to the Chesapeake Bay, from where he can get most easily to his office. Kazimir proposes New London, where he studied or, if necessary, New York, where his family awaits him. Ken takes a longer view and proposes with great seriousness that we disembark in London, England. As for me, I incline toward Boston, whose longitude we have just crossed—Boston, the birthplace of Franklin. However, ten more days are yet to pass, and who knows where we will be then.

Life aboard continues with its big problems and small controversies. For example, we note that, between 10 a.m. and 3 p.m., for the same depth (152 meters) the temperature of the water has risen from 18.75° to 19.14°C.; why this increase in temperature at a constant depth? The mesoscaph will warm up a little, and with a certain delay due to its inertia it will probably rise several meters—when the water will already have grown cold. Then obviously we will not understand why. Assuredly our computer would render us great service.

Another odd phenomenon is the attitude of the mesoscaph in the horizontal plane: in general it advances with the conning tower toward the rear, but it also frequently oscillates in this plane almost 180° in one direction or the other. This oscillation takes place over a period of several hours so that it does not disturb us at all. Occasionally, too, the mesoscaph makes a complete turn. If our count is correct it has now turned all the way around eight times since the beginning of the expedition.

Our speed continues at the same rhythm; on August 5 at 1 a.m. we record an average of 3 knots for twenty-four hours. We have moved well toward the north; to within a narrow margin we are inside the average course of the Gulf Stream for this time of year. According to the *Lynch* the Gulf Stream at this point has a width of more than 100 kilometers.

55

THE VISIT OF THE TUNAS

Since the start of this dive I have been hoping—indeed I considered it almost inevitable—that we would see shoals of fish almost steadily moving around us, as they so often do when a vessel is drifting on the surface. We even discussed what should be done if there were so many fish that the portholes were obscured by them—in other words, if we could not see the forest for the trees. This nearly happened with the *Auguste Piccard*. After some days idling on the surface in the port of Vidy (Lausanne), we were gradually surrounded by so many small fish that it became dark within the mesoscaph and many of the portholes were completely obscured. I was advised, if this happened in the Gulf Stream, to use flashbulbs to drive off the fish, most fish being disagreeably affected by sudden change in light.

However, up to now, unhappily, not a single shoal of fish has been seen, only a few isolated individuals, and these but rarely. Now finally, on August 5, at a depth of 200 meters, a fine shoal of big fish swims by at some distance above us. At first we cannot identify them; from time to time they approach, withdraw, then return again. For almost a day and a half they remain with us. At least we think they are the same, although possibly there are several similar shoals. Each time they come close enough we try to seize on some detail that will permit us to identify them. Finally we agree: they are undoubtedly tuna; we incline to think bluefin tuna (*Thunnus thynnus*), the famous "red" tuna of the French. The top of their bodies is dark blue, the underside light in color, which supposedly makes them less visible to their adversaries who, if they attack from above, see them against the dark background of the depths, or, if from below, see them hardly better against the light background of the surface.

However, when the photographs were later examined by a specialist, Philippe Serene of the Centre National d'Exploitation des Océans in Paris, the long pectoral fins caused him to identify them positively as *Thunnus alalunga,* known in France also as white tuna and in England as longfin tuna, or sometimes incorrectly as albacore. Many varieties of these tunas, in particular the red tuna, the white tuna, the patudo or tropical tuna, and the albacore, are much alike and often hard to differentiate. All the tunas, belonging to the scombrid family, are close relatives of the mackerel, the swordfish, the bonitas, and even the superb sailfishes.

It is interesting that they are only found, at least according to certain authors, in waters with a temperature between 14° and 24°C. and salinity of 3.5 percent. Their blood, not actually warm, is nevertheless 8° above that of the surrounding water, and this characteristic doubtless contributes to their indomitable energy. A glance at our thermometer shows that the sea is 18.13°C., which is in conformity with the rules for tunas. I do not have before me a record of the exact salinity of the water, but I know that it is in the neighborhood of 3.5 percent.

These tunas, which are at least half a meter to a meter in length, with some perhaps exceeding a meter, are swimming about the mesoscaph rapidly and supplely, sometimes approaching close, sometimes 10 to 20 meters above us. We see them through all the portholes in succession. They seem not to be attracted by the light, at least not so much as the plankton, but at moments they appear to show signs of agitation. Is this when we turn our searchlights on or off? We are not sure. Presently we are all at the portholes alerting one another:

"Look out, they're coming your way!"

The cameras begin to click, with no great results, by the way, for the tunas go by too fast and never at the right distance.

On the other hand, when we completely blew out one of the VBTs, thus producing a great column of air rising from the mesoscaph, Erwin was able to see the school of tunas rush into the column, obviously reveling long and avidly in this unexpected bubble bath.

It appears to have been the Abbé Bonnaterre, one of the Encyclopedists, who in 1788 published the first description of the white tuna. Later it was established that, imitating the eels in this respect, tunas prefer the Sargasso Sea for their breeding place. Where do they go from there? They have been found practically everywhere in the Atlantic and especially in the equatorial Atlantic; but they systematically avoid cold waters. Like their cousins the red tunas, were they crossing the Atlantic, and like us were they making use of the Gulf Stream to travel faster? That could save them several days from one coast to the other. But why in the world are they in such a hurry?

For us this visit of the tunas was absolutely fascinating. Moreover, it opens new horizons for commercial fishing. To my knowledge no one had ever before observed a shoal of tuna from the depths. No one had ever been able to note their disposition, their reactions, their gregarious habits, their speed, all of which are so easy for us to study through our portholes. Deep-sea fishing, for tuna in particular, is still done by primitive and empirical methods. Many research laboratories dream of using submarines to arrive at more rational methods of fishing. Such specialized submarines would be costly, and few people really understand the advantages to be derived from them, the extraordinary investment they can represent. We, comfortably watching through our portholes, can in the course of a whole day or even a day and a half, observe our tunas in complete tranquillity. What a pity we do not have a tuna fisherman with us, or else a real icthyologist.

According to statistics, in 1968 a total of 60 million tons of fish was caught throughout the world. It has been estimated that simply by fishing in a rational way one could at least triple this catch, and without risk of exceeding the limit that nature authorizes, without endangering the species.

56

THE VISIT OF EL COYOTE

On August 6, we employ again, and again in vain, the decontaminator that is supposed to destroy carbon monoxide; no result. The proportion has now reached 30 parts per million, already above the 25 fixed at the start as the absolute limit. Oddly enough, we have on board an adventure novel, *Ice Station Zebra* by Alistair MacLean. Busby, who is reading it with interest, happens upon a passage describing life aboard a nuclear submarine, wherein the author says that 30 percent carbon monoxide is absolutely normal!

August 7. El Coyote, the airplane which I have mentioned, flies over us, crossing and recrossing the Gulf Stream to determine its exact course and confirm the observations of the *Lynch.* With a little luck we could almost see the bathythermographs it drops passing before our portholes. Despite the remarkable field of vision we enjoy (there is, in fact, a place from which one can view the sea through twelve different portholes without changing position), it does not extend far enough, and we see none of the instruments dropped by the plane. Our navigation continues to be good and for the last forty-eight hours we have followed exactly the average course of the Gulf Stream.

During the afternoon I propose a new descent to 500 meters; we could, as an experiment, remain just four hours and have the surface determine our position at departure and return, thus making it possible to calculate the precise speed of the current at that depth. This proposal does not meet with Kazimir's approval, for it had not been planned for the mission. He rallies, however, in view of the ease of the operation and the interesting results that can be hoped for, and so, with Frank, we make preparations for that night. Our speed will probably not be very different, taking into account that here, at almost 300 meters down, the current speed in relation to the surface speed appears to be only 0.2 knot.

In general, moreover, the current is much more uniform now than at the beginning of the drift. The internal waves are also much less marked. At the last moment we have to cancel the descent, at least for tonight, because a sailor on the *Lynch* has just had a heart attack and the vessel must get him to shore at top speed. The case is fortunately not serious, according to Dr. Jessup; a Coast Guard vessel will meet the *Lynch* and take charge of the seaman with the utmost care. So our descent cannot take place until the night of August 9-10 on the return of the *Lynch,* which is especially well equipped for precise navigation.

Next day, Monday, August 8, our speed is recorded at 3.3 knots for the last twelve hours. The surface is calm, the state of the sea at zero. Whales and sharks have been sighted; Cyrus Epler, captain of the *Privateer,* has caught a shark more than a meter and a half long.

57

TOWARD THE BERMUDAS

Hurricane Anna has vanished. We are so far to the east that we are moving off my chart. We have only one chart left, and if we continue at this speed in this direction, in three days it will be no good to us.

We are now deviating from what is called the average trajectory of the Gulf Stream by "redescending" directly toward the southeast. This average trajectory, however, means nothing in this connection, and of course we are staying in the Gulf Stream by following its actual course. However, at the end of the afternoon we discover that we are closer to the Bermudas than to New London. There is no danger that we will go as far as those islands, however attractive they may be, for the Gulf Stream swings in a wide circle around them, sometimes taking them as a center, without ever reaching them. The Bermudas and the Sargasso Sea are unacquainted with this current.

Beneath us the sea is very deep—more than 5,000 meters at present. When we talk on the telephone, whole sentences are thrown back at us by the echo, often much clearer than the original message. On board life continues to be calm; right now everyone is asleep except Ken and myself. We are at 276 meters; for Ken the French Revolution has ended (17.93°C.), the salinity of the water is at 36.10 per thousand, and the speed of sound is 1,521.9 meters per second. The humidity remains bearable; more than a hundred sacks of silica gel are hanging from the ceiling like bats in a barn.

Next day, to celebrate our last week end aboard, we glide at 4 knots in an east-southeasterly direction, straight toward the open sea, without knowing what port we might one day arrive at.

Whatever our direction and our speed, followed by the acoustic eye of our escort, we continue our work and our daily routine. Today we are once more changing the panels of lithium hydroxide, for the proportion of carbon dioxide has reached almost 1.5 percent, the extreme permissible limit. Actually, I do not entirely approve of this method. Since the lithium hydroxide we have aboard does not allow the carbon dioxide in our atmosphere to exceed 1.5 percent, this means that it could absorb *all* that we produce. If it can absorb all, why let the percentage mount as high as 1.5 when obviously 1 percent is more than enough? A better efficiency of lithium hydroxide needs to be worked out.

The answer that I am given—that the panels are barely able to keep the carbon dioxide within the limit of 1.5 percent, meaning that they become saturated—

naturally does not satisfy me; it simply means that our panels are not arranged in an optimal fashion and that we should improve the arrangement. Also the bags of silica gel are too big, which results in low efficiency; the same quantity of silica gel properly distributed could have produced a relative humidity of between 60 and 70 percent, instead of 70 to 80 percent, as now. But, everything considered, our interior atmosphere remains fairly comfortable, except that the temperature could preferably be a few degrees warmer.

Finally on the evening of August 9, we make another deep descent; I want especially to check the speed of the current at 500 meters depth, as I have said, and Frank and Ken will profit by the opportunity of having a few more "caps" and "suss" set off. The descent proceeds exactly according to plan, showing, first, that the sea is uniform here "according to the book," and, second, we are beginning to understand our vessel thoroughly, to know how to operate it unerringly, even in relatively ticklish situations.

An hour and a half after our departure we are stabilized at 500 meters and the *Lynch* gives us our precise position established through Loran C. For four hours the *Ben Franklin* remains at this depth, not varying in altitude more than its own height.

The explosions, on the other hand, are strong, in fact too strong. This is the occasion, mentioned earlier, when a small object fell from one of our shelves, and we felt the entire hull vibrate under the shock of successive waves. We ask the *Lynch* to go a little farther away, and it takes up a position 2.5 kilometers from the spot directly above us.

The surface informs us that we will leave the *Ben Franklin,* the expedition terminated, on the morning of August 14 and not on the evening of August 13, as "we have a right to do." Personally I agree with this decision, but there is a veritable explosion of fury from some of my companions, not because of the chosen date but because it is imposed without giving us a chance to discuss the matter. Decidedly the surface, always well-intentioned, has not yet found a way of making the orders received, or the instructions it considers best, conform with the team spirit that has developed among us. In any case, since the decision "imposed" is an expression of common sense, the resentment is short-lived, and pretty soon our usual pleasantries are resumed.

"*Ben Franklin, Ben Franklin.* Over."

"*Privateer,* this is *Ben Franklin.* Over."

"*Ben Franklin,* is Busby there?"

To this preposterous question we give the appropriate answer.

"*Privateer,* this is *Ben Franklin.* Frank Busby is not here. He has just left and we have no idea when he will be back. . . ."

At 12:40 a.m. just before we reascend toward the 200-meter depth, the *Lynch* once more calculates our position and consequently our speed: at 500 meters the Gulf Stream here is still moving at exactly 3 knots. A rapid calculation shows that the mass of water moving with us is at least 75,000,000 cubic meters a second, and probably much more since the current here may descend ten times deeper than the point where we are now.

At 8 a.m. on Sunday, August 10, we are well balanced at 160 meters, drifting in water at 19.58°C., and the first unusual thing I see is a vast cloud of little greenish-brown particles surrounding us in every direction with a density of about 200 to 400 particles per cubic meter. Are these algae (quite possible at this depth), some of the phytoplankton I have already mentioned, or miscellaneous debris perhaps from a great distance? Frank, to whom I put this question but who is essentially a geologist, considers it of no consequence whatever.

58

ON THE EDGE OF THE CHART

On August 10 at 10 a.m. we are at 151 meters and in a gently ascending current. Three quarters of an hour later we are at 142 meters. At moments we hear the gas escaping from the batteries and thus changing our weight again. It is absolutely necessary to descend a little. However, one of our VBTs is full and the other contains, since our last descent, air at 50 atmospheres. To make water enter at 14 atmospheres we must first get rid of this excess pressure. Since the VBT has a capacity of roughly 350 liters, it contains at 50 atmospheres some 17,500 liters of air, in the neighborhood of 23 kilograms. To let the water come in, we must overcome the difference, that is, 350 liters at an average pressure of 35 kilograms per square centimeter, representing a little more than 16 kilograms of air. The situation is paradoxical: before being able to take on weight in order to descend, we must lose 16 kilograms of weight (which on our scale is enormous), and this is the only way to get water into the VBT. The mesoscaph takes advantage of this operation to rise to 110 meters below the surface. A few moments later having once more taken on sufficient water, it begins to descend again and proceeds to stabilize itself at first at 140 meters, still with a certain downward tendency because we are now following a descending branch of the Gulf Stream. What is much more curious is the speed and the direction we maintain: an average of 3 to 4 knots straight toward the east. Tomorrow, unless something happens to prevent it, we are likely to go right off the last chart we have aboard.

The only really important thing is that we continue to remain well placed in the Gulf Stream, halfway between the north wall and the center of the current, if not the center from the point of view of speed, at least insofar as temperature is concerned. Thus in all probability we shall continue to drift in the current until August 14.

August 11. The sea is becoming rough again; a storm is brewing and the two surface vessels have trouble taking their bearings. The Loran system of which I have spoken is remarkably precise when atmospheric and ionospheric conditions are favorable, but difficult to interpret in stormy weather. Even the *Lynch,* an Agor (Auxiliary General Oceanographic Research) ship, well equipped with Loran A and C, is no longer adequate in this situation. It has not yet been fitted out for navigation by satellite. This equipment is now on the market, and the navigation satellites orbiting the earth allow incredible precision. Not only can location be determined through the Doppler effect, in relation to the satellites, from which the position is known from tables published regularly, but these satellites are so kindly disposed that periodically they themselves announce exactly where they are so that necessary corrections can be made. These corrections obviously are calculated on the ground and transmitted to the satellite, which repeats them on demand.

Nevertheless, to our surprise, in setting down our position, which we fear will have to be inscribed on the wall of the mesoscaph—for at this speed we should have sailed off the chart—we discover that our heading has changed completely and that we are now proceeding toward the north-northeast. And so our position is still barely on the chart.

59

FIVE O'CLOCK TEA AND HURRICANES

Tonight we are going to make our next-to-last descent, to 500 meters. First, a noteworthy event takes place aboard: Ken Haigh, to warm us up, makes genuine tea —tea in the true English tradition. Don Kazimir lends him a small electric kettle and, what is more essential, 125 watt-hours to get the water boiling. Ken produces a tin of Darjeeling from under his berth, clandestinely brought aboard, along with a Sèvres teapot. With the silent concentration that always attends the preparation of tea in Great Britain, he measures out the tea leaves, estimates the right quantity of milk and sugar, and finally offers each of us a beverage the like of which we have not tasted in weeks, a veritable nectar which has an indisputable advantage over that of the gods in being real.

Then, after a rather unnerving communication the surface—a new hurricane has been reported; how much time we would need to reach the surface if that should be necessary?—and a sympathetic though one-way conversation with a new troop of porpoises, we can begin our descent.

Erwin is at the pilot's station. At 8 p.m. he opens one of the VBTs. Two hours later we are stabilized at 500 meters: the *Lynch* can explode its grenades; this time it is Frank Busby who asks the *Lynch* to remain at least 3 to 4 kilometers from the point directly above us. It would be too bad to be blown up two days before the end of our mission.

All evening our mood is good. We have almost reached the long vacation. That new tropical depression, in fact, is the one thing that bothers us. Like those of all "depressed" creatures, its reactions are unpredictable, but though we naturally have to be on our guard, there is no reason for the moment to be unduly concerned. At five minutes past midnight we begin to ascend again and at 7:00 a.m. on August 12 we are once more in good equilibrium at 200 meters. The temperature—18.55°C.—is exactly what it was at the same depth twelve hours ago before the beginning of the descent. We are certainly in the Gulf Stream, but our position continues to surprise us: we are now bound due north, headed straight for Nova Scotia.

Once more the weather is bad on the surface; Bill Rand tells me by telephone that the swell has risen nearly 8 meters, along with choppy seas driven and torn by the wind.

This is not a "hurricane." It is a heavy sea. The hurricane itself, it seems, is nearly 400 kilometers distant. If we had to surface in this weather, tie up the mesoscaph to the *Privateer,* and begin a tow, I believe we would have a difficult time. After all, the weather may improve during the next twenty-four hours (or it can get even worse). On the other hand, we encountered weather just like this at Guam in 1960 when we came up from 11,000 meters, and we got out of it all right, but on that occasion we had a powerful destroyer as surface escort and a seagoing tug with all the necessary equipment for an operation of that kind. In the midst of the tempest raging on the surface, a piece of good news gets through to us. A Coast Guard vessel, the *Cook Inlet,* from one of the so-called permanent North Atlantic stations that serve as meteorological bases and also as reference and rescue points for transatlantic planes and ships, is to be relieved at station "Echo" and will be passing within a few hours' sailing distance of our probable position when we come to the surface day after tomorrow.

This news cheers us immensely. For one thing, there is a good chance that the *Cook Inlet,* a comfortable cutter, will take us aboard and transport us to land in about a day and a half—surely the quickest way of getting there. Upon reaching the surface, all our notes and equipment will have to be transferred and the mesoscaph taken in tow by the *Privateer* regardless of weather conditions. The prospect of assistance from the *Cook Inlet* is most welcome, for we know that the Coast Guard is trained to meet every trial and hazard, capable of rescue performance in any weather, on all the seas of the world.

I have always felt great sympathy for the men of the Coast Guard, whose mission demands foresight (they establish the safety rules of the sea), surveillance, lifesaving skill, and sacrifice (for the sea levies a heavy toll on them too, despite their courage and competence).

Now our own position is no longer our principal interest (oddly enough, we continue to move due north, no doubt so that we will not sail off our last chart); it is that of the *Cook Inlet.* At the end of the afternoon on August 12 the *Cook Inlet* is 300 kilometers east of us. Her speed being some 14 knots, she will certainly arrive at the appointed spot before us. However, we already know she will wait for us, and we are grateful to these sailors who, although anxious to return to their home port and see their families again, are willing to pause and take us aboard.

At 7:45 p.m. we are at what might be called the twenty-fifth hour. A Strauss waltz booms from the cassette player and we are about to descend for the last time, to bid final adieu to the depths at 500 meters. Erwin is at the pilot's station. This descent poses no further ticklish problems for him, but of course it demands close attention. We also know now that even if we lose the Gulf Stream it will not be very important. We have in any case accumulated so much data that it will take months to analyze it.

60

LAST DESCENT

We continue to glide at 3–4 knots toward the north. The descent lasts two hours and we remain only one hour at maximum depth, the time necessary to make some acoustic measurements above a bottom of 5,000 meters. The temperature of the water here remained at 12.59°C. as against approximately 6.5° for the same depth off Palm Beach (500 meters), allowing the supposition that the Gulf Stream goes even deeper.

The ascent is particularly splendid. One of our most powerful searchlights is turned on, using 1,000 watts but producing three times the light of an ordinary bulb of this power. We rise gently, allowing plenty of time for the plankton to accompany us. For this final trip the sea again affords a magnificent spectacle. Salpa dances, nebula ballets, jellyfish, necklaces of gold and silver, these jewels recalling the art of Mycenae or Etruria, all accompany us during an ascent of 300 meters, swimming tranquilly, undulating, gliding around the mesoscaph, passing our portholes again and again in a fairyland setting. Even a shark comes and pauses briefly in front of my porthole, but seeing nothing that interests him vanishes at once into the blackness from which he came. He is probably a blue shark, the *Prionarce glauca,* especially slim and delicate, nose sharply pointed, pectoral fins long; he lives chiefly in the region we are now crossing and is at home in waters from 7° to 21°C. (at that time the water around us was about 19°C.).

Now our final day begins. This is August 13. We decide to let the *Ben Franklin* oscillate all day between 200 and 300 meters. We commence to prepare for our emergence tomorrow, with undeniable pleasure but not without a certain melancholy as well. We have become so accustomed to this life. Everything is so comfortable aboard, except for some minor problems with the life-support system on this last day. We have now reached 40 parts per million of carbon monoxide, which is the standard limit, not to be exceeded. Some of the pipes also present problems. But on the whole everything is going admirably. We are used to the temperature too (19°C.), and to the humidity, 73 percent at the moment. The interior pressure has slowly increased despite the output of oxygen which has been kept within normal limits and the excellent absorption of carbon dioxide. This slight increase in pressure is due to a pressure reducer which was not completely airtight at the start. As soon as we noticed it we kept it in its normal closed position and have not opened it except for brief moments as needed. This interior pressure, 1.12 kilograms per square centimeter as against 1.01 at our departure, which would correspond in the case of a dive to a depth of 1 meter in open water, is insignificant, even for a period of several weeks.

61

DEEP SCATTERING LAYER

At 10 a.m. on August 13 comes a telephone call. The *Atlantis II,* the ultramodern oceanographic ship from Woods Hole, is passing directly overhead and asks if she can do anything for us: she is especially well provided with acoustic instruments, and certain common measurements might be interesting. We ask immediately if they have seen any traces of the Deep Scattering Layer (DSL) in this region; unfortunately the reply is negative. Dr. Fuglister, one of the Gulf Stream specialists whom I had met thirteen years before, is on board. However, since our mission is practically over, we merely exchange a few friendly words.

The absence of the DSL is a disappointment for me as it is for Busby and Haigh. Never has any submarine, never has any oceanographic team been so favorably placed to study this layer. These layers are more or less mysterious; it is generally known that they reflect sonic and ultrasonic waves sent out from the surface and therefore lead people to believe that the bottom is much closer than it actually is.

Oddly enough, the layers are not at a constant depth; not only do they vary in different oceans, according to the season and other factors that seem to us irregular because we do not understand them, but the depth of a layer varies almost systematically with the hour of the day and night. During the night it is close to the surface. At dawn it descends and "takes refuge," usually between 400 and 800 meters down.

Countless scientific observers have devoted themselves to this problem. At first it was believed that these were layers of plankton, perhaps made up of the euphausids, mentioned before, but more precise experiments have not entirely supported this hypothesis. Now it is believed that these layers are of different origins, sometimes made up of minute plankton and sometimes of fish, in particular Myctophidae, or of both. The Myctophidae, a suborder of the Bathymalacoptera, which are also called lantern fish, exist in great numbers in the ocean. At least a hundred species are known. Small in size (5 to 15 centimeters), most of them are probably luminous, having a number of "lanterns" or photophores on their bodies and tails. It is known that by day they live about 800 meters deep and rise toward the surface at nightfall.

We had hoped to spend hours, days really, in these reflective layers, to follow them as they rose and fell, to make still and motion pictures, to try to determine which acoustic waves penetrated them best and which waves were reflected by them. But alas, on the thirtieth day of our mission, not the *Ben Franklin* or the *Privateer* or the *Lynch,* and now not even the *Atlantis II* has seen the slightest trace of DSL in our neighborhood. Our special acoustic equipment, our cameras, our stability, and our extreme maneuverability have not served in this endeavor, although happily we have something in store for a future campaign.

The day is peaceful. Our group has transformed itself into a team of housekeepers. We considered it important to bring to the surface a mesoscaph impeccably clean and neat. Everything must be carefully stowed. In a commercial vessel this is the responsibility of the first officer, but since we have no first officer, everyone takes part.

The work is easy, for we have kept the mesoscaph in perfect condition anyway. Now it must be ready for a long tow, more than 1,000 kilometers to New York, if that is to be the port for the *Ben Franklin,* and this in any kind of weather. Today, however, the sea is better; at last report at 8 p.m. the state of the surface is at number 2 (Beaufort's scale), the wind is blowing only 10 to 15 knots, and there are no clouds. We are 800 kilometers from Portland, Maine, where no doubt the *Cook Inlet* will put the crew ashore. At 8 p.m. we hear the news, the "Voice of the Gulf Stream" which comes to us for the last time. For the last time too Chet May puts "The Yellow Submarine" on the cassette player.

We cannot keep from looking repeatedly at the clock and announcing different anniversaries:

8:25 p.m.: it is one month since the drift began.
8:34 p.m.: it is one month since we closed the hatch.

8:54 p.m.: it is one month since the mesoscaph was totally submerged off Palm Beach.

The last night is calm. Those who were asleep woke up early. It has been decided that we would emerge at 8:00 a.m. on August 14. The ascent (we were at 300 meters at 1:15 a.m.) must be gradual to allow the gas from the batteries to escape gently without carrying electrolyte or oil with it. Also we must wrap up our notes, documents, records, films, cameras all very carefully in plastic bags, of which, by the way, we had an incredible profusion aboard. Actually, food, clothing, bedding, and many other accessories had been wrapped in these plastic bags at departure—was this what finally produced that excess of carbon monoxide?

Everything is ready well beforehand, stacked in the forward hemisphere ready to be transferred to the boat that will come to get us. The round table in the "wardroom" is dismounted and stowed; the chairs are fastened against the wall; finally, the ladder that has been suspended against the ceiling for thirty days is taken down and put in place beneath the hatch which tomorrow will open into the conning tower.

62

FINAL ASCENT

At 1:15 a.m. Erwin Aebersold blows air into the starboard VBT for ten seconds.

Very gently the *Ben Franklin* begins to move toward the surface.

4:32 a.m.: another four seconds of air.
5:00 a.m.: we are at 260 meters.
6:50 a.m.: we are at 150 meters.

We begin to hear the first gurgling in the batteries. Soft music continues to come from the cassette player; Frank and Chet are singing a sad sort of melody. This is not habitual with them, but they seem a little melancholy this morning. That's natural enough, of course; in a few hours we are going to abandon our house, the habitation in which we have passed, at the mercy of a vagrant current, a thousandth part of our terrestrial existence. It is said that prisoners have abandoned attempts at escape after taking a last backward glance at the cell they have done everything to abandon!

7:12 a.m.: it is now quite light, we are at 93 meters. The ascent has been steady, but here we encounter a layer of warmer water and bounce downward a few meters. No matter; Erwin blows a little more air into one of the VBTs. In any case we still have a large reserve of available air.

7:44 a.m.: the surface is not yet visible. The water is at 23.22°C. I cannot give a date in history to correspond with that temperature.

A minute later, at 36.50 meters depth, the surface of the water appears clearly. We do not see any whitecaps but there are waves which we begin to feel at a depth of 20 meters, indicating a considerable swell at least.

A small jellyfish—is this the last denizen of the sea we will encounter?—passes tranquilly in front of a porthole. Through the same porthole I see the stormy surface onto which we will break in a few minutes.

The appearance of that surface seen from below is awe-inspiring. The sky is not visible. Eddies and whirlpools, alternating with areas of absolute calm, form huge circles and infinitely varied geometric figures above us. It is an unforgettable spectacle to see oneself approaching, decimeter by decimeter, centimeter after centimeter, to that far from welcoming surface which is nevertheless the only way to terra firma.

Just before we break the surface we see once

more some of our friends the salpas, completely transparent in the clear, luminous water at 9 meters depth. They approach to gambol for a last time and make their farewells.

63

THE EXPEDITION ENDS

At 7:57 a.m. the antenna emerges. Radio communication is quickly established. Of course during the whole ascent we have been in contact by underwater telephone.

At 7:59 Erwin opens the valves to empty the main ballast tanks; in a few moments the *Ben Franklin* appears boldly on the surface, and its gentle rolling tells us at once that the sea is not as bad as we had feared.

It is 8 o'clock on the morning of August 14, 1969. The Gulf Stream Drift Mission is over. We can open the hatch.

Because of the slight excess pressure inside we have to open the hatch gently. As a matter of fact, there is an equipression valve that can be used in these cases; it allows the interior pressure to be equalized gradually with that outside, in one direction or the other. For some reason Don Kazimir, a bit impatient, no doubt (which is understandable), does not like to use this valve, and so we must open the hatch very carefully at first. There is a slight whistling in our ears. At 8:09 the hatch is finally wide open.

Don Kazimir courteously suggests that I emerge first. I refuse this honor and pleasure and, on the contrary, propose that the guests leave first; I still consider myself in my own house here. Ken Haigh, Chet May, and Frank Busby go on the bridge; next, Erwin Aebersold and Don Kazimir, and finally myself.

On the bridge! We are on the bridge. In the open air. Our faces are instantly swept by wind and spray. Already on our bridge are members of the crew of the *Privateer* and the *Cook Inlet.* I pass over the "How are you?" and the "My, you're looking well!" These have no importance. The sea is there, the boundless surface, rugged without being tumultuous. The sky is gray but we think nevertheless that we see the sun—that sun which accompanied us throughout the dive—and around us is a whole flotilla.

First the *Privateer* that we know so well, a little farther off the *Lynch,* all white, which we see for the first time; The *Cook Inlet,* white too, striped with red like all Coast Guard vessels; the *Atlantis II* of Woods Hole which has returned to this area to see us emerge and to offer its services if needed; a number of rubber boats bobbing like corks as they make their way up the waves, heading toward us. The navigation and tracking have been so precise that the *Ben Franklin* could emerge in the midst of this flotilla without risk of colliding with anything. A few hours before, another vessel had been there: a Russian "trawler" fishing for tuna or shrimp, which in passing perhaps listened to us in silence.

Rapidly the baggage is brought out on the bridge. A rubber boat is tied to the *Ben Franklin* and *all* our notes, *all* our documents, *all* our work for a whole month of research is heaped in this frail craft. I propose that at least two trips be made so as not to "put all our eggs in one basket." The reply I receive, which is logic itself, is that "this would double the risk of losing half of them."

The sea is rough, to be sure, but our sailors know their business. The boat arrives alongside the *Cook Inlet,* which manages a half turn in place in order to provide us with a lee, and without any trouble we climb aboard, one after the other, our equipment following without loss, without even receiving a fleck of water.

Our reception aboard the *Cook Inlet* deserves special mention. Captain Richard K. Simmonds welcomes us like princes or admirals. The best cabins are assigned to us; the crew will crowd themselves into the less desirable quarters. Everyone outdoes himself in prodigies of kindness, which seem altogether natural to these good-hearted men, to make us comfortable during the thirty-six hours we will spend aboard, and indeed this transition period between our 732-hour dive and the hectic life that awaits us ashore was extraordinarily beneficial for all of us.

It was interesting, too, for the *Cook Inlet* is practically an oceanographic vessel, especially equipped as a meteorological station with sounding balloons and radar. We sped at 14 knots toward Portland, Maine, arriving on the afternoon of August 15. A special Grumman airplane then took us to Bethpage.

Meanwhile the *Ben Franklin,* towed by the *Priva-*

teer, was also proceeding toward Long Island; it reached there about a week later. After several days for cleaning and checking, the mesoscaph was again taken in tow by the *Privateer.* This time its destination was New York harbor. The crew, aboard a Coast Guard cutter, and our families and the press photographers, aboard a tug, met the *Ben Franklin* at the foot of the Statue of Liberty. There the crew transferred to the deck of the submarine. The *Privateer,* a Grumman motor launch, two more tugs, and a fireboat were also assembled there.

Suddenly into the brilliantly sunlit sky the fireboat sent up enormous jets of water in New York's traditional welcome to a vessel returning in triumph. The fountain salute continued for fifteen minutes, then the *Ben Franklin* was moored at a wharf, and the expedition came to an official end.

64

AFTERWORD

Between the invention of the bathyscaph by Dr. Auguste Piccard and the exploration of the Gulf Stream with the *Ben Franklin* sixty-five years passed, but between the first dive of the bathyscaph and our expedition only twenty-two years elapsed. Much skepticism had to be overcome; many doubts and considerable sarcasm greeted the first trials. But now great rewards in satisfaction are to be found in the more than one hundred research submarines operating around the world, the hundreds of portholes through which the ocean depths are being examined, the dozens of underwater laboratories silently gathering the knowledge on which the very future of humanity may depend.

Great nations and small have turned their attention to oceanography. The United States very quickly assumed a leading role. Its submarine output ranges from nuclear submarines, touring the world under water, passing beneath polar pack ice, disappearing for months at a time—to the small manned capsules whose miniaturized laboratories contain a whole world of electronics, and which at the end of a few hours must surface for air. Between these is a whole armada of vessels, large and small, made of various materials and for varying depths, planned for one observer or for many. Some are designed for pure research, others for the beginnings of industrial exploitation, and still others for strictly humanitarian purposes, such as rescuing the crews of navy submarines.

Under pressure from her navy, and within the framework of pure scientific research, France, since the end of World War Two, has resumed and pursued the work of Dr. Auguste Piccard. The FNRS 3 and the *Archimede* in particular have been used for a long series of scientific investigations of great value in the Mediterranean, the Atlantic, and the Pacific. The submersibles of Commander Jacques Cousteau have dived with increasing success in most of the seas of the world, and some have made descents in South American lakes. The results, projected on television screens, have shown millions of viewers throughout the world the formidable influence that the sea has on man.

Japan, the USSR, and Great Britain also possess several research submarines apiece. Even Switzerland, 200 kilometers away from the ocean, has become fascinated by a realm hitherto scarcely known. As we have seen, not only the bathyscaph *Trieste* but the mesoscaphs *Auguste Piccard* and *Ben Franklin* originated there; all three were later acquired by the United States. In the domain of the aqualung, the Swiss mathematician Hannes Keller achieved the triumph of a 300-meter dive, using a special combination of gases and accelerated decompression.

Yet if the conquest of the sea is a sign of mankind's progress, this same progress has had other results. The population explosion and the world-wide proliferation of technology are presenting a tremendous danger—the rupture of ecological systems, with the possibility of suffocating life in the sea as well as on the land. To be constructive, technological development and the exploitation of natural resources must now be directed toward the world as a single unit, with only one ecological system. Earth, with its population of nearly 4 billion, is a small, self-sufficient capsule traveling in space. If we want it to continue its journey without mishap, we must keep it clean and livable. As regards the oceans, this means that we must give even more consideration to what we put into them than to the wealth their exploitation affords.

APPENDIX

Results of the Gulf Stream Drift Mission

As these words are being written, the computers in Washington are still busy interpreting our magnetic tapes. The NASA psychologists have not completed their analysis of some 65,000 photographs made by the automatic cameras that surveyed the interior of the *Ben Franklin.* The oceanographers are still puzzling over many things that happened during the dive. Why were we once "expelled" from the current? Why only once? What creates internal waves? What prompted the attack by the swordfish? How account for the speed of the Gulf Stream, which was nothing like what we expected?

All this will be elucidated, partly at least, in the course of time. The interpretation of the scientific results of an expedition generally takes longer than the expedition itself. Recently, I visited Professor N. B. Marshall in his laboratory at the British Museum. When I entered he was absorbed in a study of a fish caught during the first expedition of the HMS *Challenger,* which traversed the oceans of the world between 1870 and 1876.

It is desirable to separate, insofar as possible, the technical from the scientific results, but this is not always easy, for the two are interdependent. Among the technical results those that concern the mesoscaph itself and those that concern the crew—that is, the life-support system—are considered separately. Among the scientific results, the observations of the oceanographers will be separated from those of NASA.

TECHNICAL FINDINGS RELATED TO THE *BEN FRANKLIN*

STABILITY

There is no need to linger over this subject. Abundant instances of the stability of the mesoscaph have been noted—how easily it can float under water for days at a time without change of depth, provided the sea itself is "stable." Obviously when the mesoscaph is caught in a system of internal waves it follows their rising and descending motion as naturally as it follows the horizontal movement of the current.

The coefficient of compressibility of the hull (35×10^{-6} cm^2/kg) compared to that of the water in the locality where we were (50×10^{-6} cm^2/kg) is altogether satisfactory. As we had foreseen, the gas in the batteries practically ceases to produce any perceptible effect beyond a depth of 100–150 meters; the thermocline which is generally found at these moderate depths tends further to neutralize this effect. The general stability was such that the reading of the measure of gravitation from moment to moment corresponded exactly with the curves that the internal waves traced on the recording manometer.

VISIBILITY

Our system of portholes had been tested, as has been seen, on a number of occasions (bathyscaphs and mesoscaphs) and nothing startling was to be expected.

On the other hand, we seriously feared, when facing the prospect of a month under water, that seaweed and barnacles would clog the portholes which, unlike the rest of the mesoscaph, could not be protected by antifouling paint.

When the mesoscaph remained in the water for several days in port at Palm Beach we had to have divers come regularly to clean the portholes. What would be the effect of a prolonged dive? In fact, we had considered several ways of dealing with the problem (underwater cleansers, special chemical protection, sending divers from the surface) but none of these had seemed feasible, and if the portholes had gradually become fouled we would no doubt have had to return to the surface to have them scoured. As it turned out, to our great relief, no trace of barnacles appeared during the whole voyage. Was this the effect of pressure? Of temperature? Or light, which at 200 meters is greatly reduced by the water? We do not yet know. However, when the mesoscaph arrived in New York after its dive, shellfish began to develop rapidly on the Plexiglas. Oddly enough, they were identified by a biologist as belonging to a species typical of Florida waters and unknown in New York. So they had made the trip with us but had obligingly waited until the end of the mission before proliferating.

BATTERIES

The batteries too behaved well throughout the dive. True, the insulation of some of the circuits could have been better, and in fact this was improved to a considerable extent after the dive. The losses occurring from faulty wire insulation were negligible; no circuit failed. In view of the novelty and relative complexity of the system we continually envisaged the loss of some groups of batteries and we economized sufficiently on current to have been able to continue the dive even without one or two of the groups. Thanks to this prudence, which actually proved excessive, we consumed only 52.1 percent of our original 756 kilowatt-hours. In a sense this was a pity, for we could have gained much by using more current, especially for our searchlights, and we could have brought back more still photographs and movies, but our basic purpose was to complete the mission and obtain the scientific results we sought. Except when we ran our motors continuously for five hours, we always remained well below the "authorized" consumption of current, and even on that day we used up only the amount we had saved during the first ten days of the dive.

The propulsion batteries were utilized only to 44.6 percent of their total capacity, the other elements for technical purposes to 64.1 percent, those designed for oceanographic measurement to 49.9 percent, and those for NASA to 80 percent.

I must add that we always took scrupulous account of the quantity of current being consumed. The automatic counters aboard were employed only during the first few days, in order to verify our method of direct measurement, which consisted simply in noting systematically the time that each machine was in use; we soon switched off the automatic system—which itself used up too much current.

PROPULSION SYSTEM

On many occasions we used our motors to orient the mesoscaph in the direction of the current just before touching down, before the guide rope could perform its function. The four motors always responded to our requirements without difficulty. We also used the motors for a trip of several hours on the occasion when the current drove us out of the principal branch of the Gulf Stream. There too the propulsion system functioned without a hitch, showing a remarkable flexibility of maneuver and efficiency.

NAVIGATION

The navigation was carried on entirely by our principal escort vessel the *Privateer,* which gave us our position by telephone several times a day.

The main problem for the *Privateer* was to keep in acoustic contact with us and to know at all times our depth and direction with respect to itself. This was accomplished chiefly by means of a 4-kilohertz "pinger" which emitted a pair of signals every two seconds. The time between the two signals varied in proportion to the depth of the *Ben Franklin,* thus giving this depth information to the *Privateer,* which followed our position on a directional hydrophone. The absolute distance could also be determined, thanks to a "transponder" of 16 kilohertz which we had aboard and which would reply to interrogation by the *Privateer.* The elapsed time between the transmission of the signal and the reply would indicate the distance between us. This system functioned faultlessly, and only once had to be supplemented by recourse to the ordinary submarine telephone. For the crew of the *Ben Franklin* all this went on for the most part automatically, but for the *Privateer* it was a major and constant duty which the crew accomplished without fail and without interruption for a month. The precision of the method obviously contributed greatly to the mission's success.

In practice, since the current generally ran faster on the surface than at a depth, the *Privateer* let itself drift backward until it was ahead of the *Ben Franklin* by some hundreds of meters. Then, getting under way against the current, it would cross above us, go beyond us by some hundreds of meters, and stop, then drift back again over our position. Thus it made its voyage stern first, sometimes moving against the current, sometimes allowing itself to drift. Twenty-eight hundred kilometers in the Atlantic going backward? Well yes, undoubtedly this too was a first.

In addition to this, the surface had to carry out the "tracking" of the principal current to determine our position in respect to the center of the Gulf Stream. For this purpose the USNS *Lynch* scoured the sea in front of the *Ben Franklin,* measuring the temperature of the water to a depth of about 500 meters by means of the expendable electric bathythermographs already described. Assuming that the maximum speed of the water corresponds with the highest temperature, which is generally the case, we could always have a good idea of our position. On a number of occasions also, the exact location of the Gulf Stream was determined by El Coyote, that specially equipped airplane belonging to Navoceano.

TECHNICAL FINDINGS RELATED TO SURVIVAL ABOARD

INTERIOR TEMPERATURE

The highest temperature aboard was 29°C., during the several hours when the mesoscaph was being towed on the surface on July 26. Slightly before that when the propulsion motors had been used continuously for several hours, a mark of 24°C. had been briefly reached. Aside from that, the interior temperature generally was maintained, when the mesoscaph was in good equilibrium and stabilized, at 1.5°C. above the temperature of the water. During the first part of the voyage, when more hours were spent at a relatively great depth, the interior temperature varied between 20°C. and 11.7°C, and we felt cold. During the second part, the interior temperature oscillated around 19°C., which was not warm but was more nearly comfortable; in any case it was bearable. But before spending several days in water at 0°C. to 10°C., it would be necessary to provide protective clothing specially designed or else to insulate the hull of the mesoscaph thermally, or both.

HUMIDITY

Thanks to the silica gel in plentiful supply, the humidity was perfectly well controlled from the first day to the last. As a general rule, we exposed the silica gel in such a way as to maintain the humidity at between 70 and 80 percent. The graphs show only five or six points beyond these limits. For this we used small bags of silica gel, the total weight amounting to about 1,100 kilograms. At the end of the mission we had half the amount left. If it had been necessary, we could have used this silica gel more efficiently (smaller bags, forced ventilation) and the same weight of silica gel would have allowed us to maintain a lower degree of humidity, although for our purpose 70–80 percent was a satisfactory proportion. On the whole, the humidity did no harm to either crew or equipment.

INTERIOR PRESSURE

In general terms the interior pressure remained constant, equal to the pressure at sea level at the moment and in the place where the dive began. In fact, the variations in temperature, the partial pressure of oxygen and carbon dioxide especially, and above all a small leak detected in one of the secondary compressed-air systems (a system which we thereafter used with great prudence and parsimony), caused negligible variations. These attained a maximum equal to the pressure of one meter of water. Without that leak, they would have played no role whatever. Its consequences, however, might have been more serious had the dive continued uninterrupted for two or three months.

CARBON DIOXIDE

The absorption of carbon dioxide produced by breathing was carried out passively by panels of lithium hydroxide (LiOH); these panels, twelve in number, were changed as soon as the carbon dioxide reached a level of 1.5 percent, a figure determined in advance—in practice, every three days. Altogether 120 trays were used, that is, some 400 kilograms of alkali with an efficiency of 75 percent. Just as with the silica gel, better utilization and increased efficiency would have made it possible, without increasing the quantity of the absorbent used, to reduce the level still further if that had been necessary. My own recommendation for other missions of this sort would be that 1 percent of carbon dioxide should not be exceeded. At the level of 1.5 percent, respiration becomes more difficult and a slight shortness of breath is felt, especially during strenuous activity. Actually, the interior atmosphere of the mesoscaph remained comfortable throughout the dive.

OXYGEN

The oxygen for breathing was supplied by two carboys perfectly insulated thermically, each containing 125 kilograms of liquid oxygen. The rate of evaporation could be regulated automatically or manually. In practice we always regulated it ourselves to save the current that would have been used by the automatic control. The oxygen was maintained at 19.5 to 22 percent, an average of about 20.9 percent, which is normal under natural conditions. Our average metabolic rate was maintained at the normal but moderate level of 2,200 kilocalories per man per day.

CONTAMINANTS

A laboratory functioning with six men in it and with a closed circuit runs the risk of producing a series of more or less dangerous or disagreeable toxins. We had on board reliable equipment for detecting the presence and quantity of all the toxic gases that might appear. Each day measurements were made of NH_3, CO, H_2S, and SO_2 with a Draeger apparatus, and tests were made once a week for twenty-eight other substances. Three harmful substances were detected:

(1) Carbon monoxide, the proportion of which in our atmosphere rose progressively from 8 parts per million on the fifth day to 40 parts per million at the end of the voyage, a proportion above what had been estimated in advance but apparently not beyond the limits of safety—which, by the way, are not yet precisely known for similar circumstances. We attempted to transform the carbon monoxide into carbon dioxide (which in turn could have been easily absorbed), but the apparatus provided for that purpose was of no effect, probably because of the high relative humidity. A part of this carbon monoxide was produced by our respiration. It is possible that the rest was produced by certain plastics we had aboard (especially electrical insulation) which may have slowly and spontaneously given off carbon monoxide.

(2) Hydrazine, in the proportion of 0.2 part per million.

(3) Acetone, in the proportion of 200 parts per million.

We have not been able to explain the presence of these last two products on board, and the hypothesis cannot be entirely dismissed that other chemical substances made our detectors register in this way. The percentage of these substances, moreover, remained constant during the whole period in which we measured them (between the eighth and the twenty-sixth day).

FOOD

In general, our provisions consisted of frozen and dehydrated foods which we had to "reconstitute" with hot water. Really hot water was almost always lacking, however, and even when it was available this form of nourishment was horribly monotonous. To be sure, as volunteers we were all prepared to submit to almost any treatment during these thirty days, but the knowledge that the number of calories, the quantity of vitamins and proteins, had been minutely calculated in advance was not enough to make us forget the normal meals of life ashore. Apart from the condensed foods, there was a quantity of preserves which most of us gladly accepted.

As for me, convinced as I am that nothing equals good natural nutrition, I sustained myself principally on dried fruits and a sort of cake made of dates, figs, prunes, raisins, apricots, pigeon peas, almonds, Brazil nuts, and coconut meat, which turned out to be excellent and extremely nutritious. A few spoonfuls a day of sunflower seeds, pumpkin seeds, and soybeans completed these menus admirably. This form of nutrition suited me perfectly; thanks to it and to the good atmosphere maintained aboard, I remained in excellent health from the first day to the last (except for the brief cold mentioned). Most of us, nevertheless, lost weight during our month under water. This had nothing to do with the quantity of nourishment available; as far as that was concerned we could have remained comfortable in isolation for at least fifteen days longer and in case of absolute necessity for as much as a month.

Some of us regularly took small quantities of seaweed prepared in advance in the form of capsules. These capsules, supplied by Atlantic & Pacific Research, North Palm Beach, Florida, contained ground blends of twenty-one different varieties of seaweed which were formulated to produce a synergistic catalytic action on the microorganism activity in the digestive tract, enabling the body to assimilate its food in a more complete and balanced way. The twenty-one varieties, including *Laminaria cloustoni, Laminaria digitata, Fucus serratas,* and *Ascophylum nodosum,* were all harvested on the Atlantic coast of Ireland and are already commercially available in many countries.

DRINKING WATER

We had two sources of water: hot and cold.

There were four tanks of hot water thermically insulated which were intended to last for the whole mission. Only one (some 250 liters) was in fact perfectly insulated. The leaks in the vacuum insulation (the thermos system) of the others had been detected before our departure but could not be repaired in time. The water in the four tanks had been heated to almost

100°C. with the electric current while we were in the harbor, and had therefore been sterilized. No chemical disinfectant had been added, but this water remained good, drinkable, and free from bacteria during the thirty days. In the last third of the mission we occasionally reheated a little water with the current from the batteries; very little, however, for we were economizing on current.

There were also four tanks of cold water, close to 1,000 liters in all. This had been disinfected on departure with iodine (7–8 parts per million), which made it practically undrinkable, so strong was the taste, and besides offered insufficient protection against bacteria. Only a few days after we left, bacteria (of the *Pseudomonas* genus) appeared, providing a second and compelling reason not to drink the cold water. As the water was still used for washing, the bacteria thereafter appeared in samples taken from our skin, but actually they caused no real inconvenience.

OCEANOGRAPHIC OBSERVATIONS

The most tangible result of the whole expedition is probably the perfecting of a new method of research and observation in the ocean. Previously large oceanographic expeditions could be carried on only from the surface, with all the inconveniences and limitations that this implies: bad weather, relative instability of the ships and therefore of the instruments, the necessity of working at a distance and in the dark. For the first time a group of observers was able to spend a month under water and carry on without interruption a great number of observations over a distance of 2,800 kilometers. The *Ben Franklin* had been fitted out almost like an oceanographic vessel, and practically all the measuring instruments used on such ships were aboard. We had the sensation while living so long in the sea of learning to know it as never before; observation of the flora and fauna was made easier by the drift of the mesoscaph which gave it the same speed as that of the water. The drift on the bottom, thanks to the guide rope which Dr. Auguste Piccard had used twenty years earlier on the first bathyscaph, also permitted perfect observation of the marine bottom for a number of kilometers of the voyage.

Since our observations were supplemented by those carried out in the conventional fashion on the surface by the oceanographic ship USNS *Lynch,* the total result of the mission was to procure extensive information about that part of the Gulf Stream.

To be sure, no sensational discovery was made, which no doubt disappointed many people. Sensational discoveries in the sea are hardly ever made, however. On the other hand, it is important to acquire scraps of information in large quantities, which taken together contribute to our general knowledge of the ocean and consequently of the globe. Our aim had been to procure as many as possible of these bits of information for oceanographers to integrate into their systems. Some years ago the sardines that had been an important source for the fisheries of California disappeared almost completely from that area. It was found that they left under the influence of the temperature of the water, which had changed by about one degree. One degree? A great discovery? A small bit of information? Let us simply say that it was minor information of considerable importance if it provides the possibility of tracking down the sardines so as to resume fishing in another area.

The principal observations included three that were somewhat perplexing: the speed of the current, our single ejection from the Gulf Stream, and the size of the internal waves. As the text shows, we made innumerable visual observations, but since visual observations are always subject to variable interpretations, these were supplemented by precise data, obtained by various automatic measuring devices. The U. S. Navy, represented by our two oceanographers, Frank Busby and Ken Haigh, was in charge of making most of these measurements, and the results and conclusions will in time be published by the U. S. Naval Oceanographic Office. In this section only the kinds of apparatus and their purposes will be described.

SPEED OF THE CURRENT

During the first part of the expedition the current was less rapid than expected and during the second part more rapid.

But take note: it is not a question of *the* Gulf Stream but only of the stream we encountered. During that first part of the dive we descended to the bottom five times and we were never in really deep water. Now it is evident that at the bottom the current is considerably slowed down by the friction of the sea floor. Above Cape Hatteras, on the other hand, where the sea was deep, the current could retain its relatively high speed at the depths at which we drifted.

Thus on the bottom off Palm Beach at a depth of 570 meters we could detect only a very feeble current which, moreover, moved at times toward the south and left no ripple marks on the bottom. A little farther north at a depth of 540 meters off Savannah, Georgia, the current attained 1.9 knots a few meters above the bottom; we made use of it to drift for an hour, well oriented on the guide rope. I have mentioned that to the northeast of Hatteras, at a depth of 500 meters, we encountered for several hours a current moving at 3 knots. It was precisely there that we had several thousand meters of water under us.

EJECTION FROM THE GULF STREAM

This event, already, described, occurred on the eleventh day of the mission. What caused it and why we were rejected only once, are partly mysteries. The "eddies" of the Gulf Stream are well known in themselves, but actually no one knows precisely what conditions do or do not produce them; in any case they cannot be foreseen. With us, the best explanation was that furnished by the Navy oceanographers, in particular by Mike Costin, who found the Gulf Stream again after a swift search over almost 280 kilometers and put us back in its center.

The *Lynch,* unfortunately, had had to leave the scene of operations to return briefly to port, so its work was interrupted for two days. As soon as it returned to its post, the observers saw that the *Ben Franklin* was not moving and was obviously out of the principal current. The Navy oceanographers then searched the region, launching dozens of bathythermographs and sounding the depths of the sea. They soon located the current, but not until much later, when they had examined all their data, were they able to reconstruct what had happened. To the southeast of the critical area the current passed close to a minor underwater ridge; the principal branch of the Gulf Stream passed to the right (east), following the topography of the bottom, but a small part of the current moved to the left of the ridge. Since we were then on the left edge of the Gulf Stream we unknowingly departed in the smaller branch which separated itself completely from the principal current.

Presently we were some 55 kilometers away. The close correspondence between the geology of the bottom and the position of the current was certainly demonstrated by our expedition. By tracing our route on a bathymetric map, one is easily convinced that the current is greatly influenced by the nature of the bottom. Apparently it is possible, at least for the depth that concerned us, that the Gulf Stream, which carried us between the twenty-fourth and the twenty-seventh day of the mission toward the southeast, took that direction only to avoid the Kelvin Mountains extending to within 200 meters of the surface. On the twenty-seventh day we moved due north once more, apparently because the mass of the current had to pass between the Kelvin Mountains and the Pablo Mountains, the other important underwater massif of this region. There would be material for a whole study in this.

INTERNAL WAVES

This phenomenon has been known and studied for a long time. Several explanations have been offered. In general, it seems that the sea must be regarded not as a homogeneous medium but as a succession of layers of different densities. Each intermediate stratum is in effect the surface of a sea or a section of sea that can become rough or smooth almost as readily as the upper surface itself. This phenomenon exists in the air as well. It is naturally more marked on the surface between the water and the atmosphere. Earlier, in the bathyscaph, I had noted these internal waves, but the stability of the mesoscaph made it an ideal platform for studying the phenomenon. Never before had I observed such large waves. Studies of the correlation between the density of the water and the temperature on the one hand and the amplitude and frequency of the internal waves on the other will undoubtedly enable oceanographers to derive new and precise information on this subject, ordinarily so difficult to observe.

THE WASP

The basic oceanographic apparatus was an automatic device called the WASP (Water Analysis Sensing Pod); every two seconds it registered the temperature of the water, its salinity, the speed of sound in water at the point in question, and the depth of the mesoscaph. This then gave us theoretically for the thirty and a half days nearly four million oceanographic measurements as a function of the depth and the hour. (As a matter of fact, a small percentage of this figure must be deducted to take account of the time during which the magnetic tapes were being changed and also for some occasional defects in the tapes themselves.) These data have been fed into a computer which is to produce all the possible or desirable curves, indicating, for example, the temperature as it varied with depth, the speed of sound as it varied with salinity, with depth, or with temperature.

PENETRATION OF LIGHT INTO WATER

All life in the ocean depends on phytoplankton which exists only because of the sunlight that penetrates the top layer of the sea. This penetration of light into water is therefore of utmost importance for the study of the productivity of the sea and the evaluation of certain of its possible changes. As early as 1957 aboard the *Trieste* with Professor Nils Jerlov, we measured the penetration of blue light (0.481 micron) off Capri where the water was particularly limpid. The *Ben Franklin* was equipped with a sensitive photomultiplier whose readings, once they have been digested by the computer, will give the values for the luminous intensity and the transmission of light in water for the vast areas where the apparatus was employed. Here too the advantage of the method of a long drift compared to a single local measurement such as had been used up to the present is glaringly apparent; errors of measurement can be compensated for, local irregularities are wiped out, and a good average value for the whole region is obtained. It has already been noted that at 600 meters depth we could still discern with the naked eye a certain amount of daylight, although its intensity was mere billionths of that which obtains on the surface.

MEASUREMENTS OF GRAVITY

The importance that geologists attach to measurements of the earth's gravity is well known; precise measurements can be made almost everywhere on land without much difficulty; but more than three quarters of the globe is covered by ocean and measurements from its surface are extremely difficult. There now exist highly stabilized devices that can measure gravity on the sea under certain conditions, but the stability of the mesoscaph was an invitation in itself to measure the earth's gravity in the open ocean. A number of measurements had been made between World Wars One and Two by the Dutch professor Vening Meiners from a conventional submarine but under many difficulties because such a submarine cannot normally stabilize its depth without recourse to motors or pumps, whose vibration interferes with the precise measurement of gravity. With the *Trieste* in 1957, in collaboration with Professor Stefano Diceglie of the Observatory at Bari, we obtained *one* measurement of high precision (with a Worden device) while the bathyscaph was resting on the bottom.

With the *Ben Franklin* the hope was to measure gravitation continuously during several hours or several days if necessary. The mesoscaph proved perfectly adapted to making such measurements. On thirteen occasions during the dive, covering over 90 kilometers in all, the gravimeter was employed for periods of one to two hours. No special gravimetric anomaly was noted, and we could even read on the magnetic tapes the effect of the internal waves!

MEASUREMENTS OF THE EARTH'S MAGNETIC FIELD

The anomalies of the magnetic field are of equal importance for the geologist. In a different discipline from ours it was by measuring such anomalies and with the aid of a computer that most of the Etruscan tombs north of Rome were located some years ago. In our case a proton magnetometer was placed at the end of a cord 60 meters long attached to the bridge of the *Ben Franklin* and suspended from a glass float far enough from the steel hull of the mesoscaph so as not to be affected by its residual magnetism. Unfortunately, our standard equipment, although it had given good results in the course of the preliminary dives, promptly refused to work: in fact, it functioned for only two hours and over a distance of 7–8 kilometers.

PHOTOGRAPHS OF THE BOTTOM

The Navy had also equipped the mesoscaph with clusters of 35- and 70-millimeter automatic stereoscopic cameras. These, placed outside the mesoscaph, could be adjusted and timed from the interior. Naturally they were all coupled to electronic flash tubes. A total of 848 pairs of photographs was made of the bottom, at five locations where the mesoscaph set down during the dive. The analysis of these photographs, along with other data, makes it possible to calculate exactly the speed of the *Ben Franklin*'s drift at these different times. It also reveals a series of details of the nature of the bottom.

ACOUSTIC MEASUREMENTS

It is probably in this department that our equipment was the most "sophisticated." Sixteen different acoustic devices were placed on the outside of the mesoscaph. Four were the sender-receivers of the four telephones; two were the sender-receivers of the fathometers; three were employed in the system of tracking and taking bearings; two were for the safety of underwater navigation, designed to warn us of possible obstacles; finally, five devices were designed exclusively for acoustic measurements, such as detecting and recording sea sounds (natural or artificial), locating the source of any underwater echoes (as in the case of the Deep Scattering Layer, for example), and examining the nature of the bottom and the sub-bottom. It is too early to give the precise results of these measurements, but I can report that 5.5 kilometers of ocean bottom were "mapped" by sonar and 1,100 explosions set off by the *Lynch* and the *Privateer* were registered and analyzed from the *Ben Franklin*.

The records on the magnetic tapes can be fed into an oscilloscope whose screen can then be photographed. In these photographs the initial shock of the explosion can usually be clearly seen, followed immediately by a vast "anomaly" formed by the bubbles of air and gas from the explosion. The echo from the bottom is equally visible and is different according to the nature of the bottom and the sub-bottom; when the sound waves return to the surface the echo is again visible, and if it has in the course of its journey crossed a zone of dispersion (Deep Scattering Layer), this zone too is clearly visible. Though no well-marked DSL was detected during the dive, the later analysis of the photographs resulted in the discovery of a small one. If we had known about it at the time we could have seen it close by. It was located at that time approximately halfway between the bottom and the surface.

The nature of the bottom was also the object of precise acoustic studies; the appearance of an acoustic photograph makes it possible to recognize in some measure the composition of the ocean floor, and for this reason the absorption of acoustic energy by the bottom was carefully measured by Ken Haigh. The energy of the impact wave can be measured from the submarine a few meters above the bottom, and the energy of the rebound is measured once more when the wave starts off again toward the surface. The difference between these two values gives the absorption by the bottom. Naturally this absorption would vary tremendously depending on whether the bottom was of sand, of mud, or of hard rock. Measurements of this sort cannot be made from the surface, for it is not possible to separate completely the degree of absorption by the water and possible dispersion layers from that of the sea bottom itself.

THE APPARATUS FOR CAPTURING LIVE PLANKTON

In Chapter 54, I described the capture of a salpa with this apparatus. In this case, the salpa was not made the object of a special study, the apparatus being on board simply for experimental purposes. The device can easily be improved and there is no doubt about its future usefulness.

THE MEASUREMENTS OF CHLOROPHYLL AND MINERALS

Walter Egan of Grumman had developed an apparatus for these measurements which has also been described earlier. Many hundreds of measurements were made at different depths, giving good results and precise information in itself, although no gross anomaly was observed.

THE NASA STUDIES

The reason that NASA took such a lively interest in the *Ben Franklin* and the Gulf Stream expedition is readily understandable. Voyages to Mars or "simply" laboratories in space orbiting around the earth for several months or years present problems much like those faced in the preparations for our expedition. NASA, therefore, principally at the instigation of Dr. Wernher von Braun, prepared a detailed program of participation. A team of specialists, presided over by Chet May, a member of the expedition and consisting of various technicians, along with observers ready to substitute for Chet May if necessary, came and installed themselves at Palm Beach several months before our departure and familiarized themselves in every way possible with our work. The NASA team, in close cooperation with Grumman, prepared a series of programs, the principal results of which are examined separately.

SLEEP

NASA has developed an extremely light and relatively comfortable headset that permits spongy electrodes to be permanently attached to certain specific points on the head, even while the subject is asleep. Thus a continuous encephalogram can be registered on a magnetic recorder. The record reveals, mainly through the waves of the brain, the movements of the eyes and the varying state of sleep and even of dreaming. Doctors have classified a sleeper's condition in six categories: stage zero, waking; stages one to four corresponding to deeper and deeper sleep; stage five corresponding to that in which the sleeper is dreaming. In the waking stage the waves are irregular, of small amplitude and relatively high frequency; at stage four they are of great amplitude and much lower frequency. It is easy to tell if the patient was reading a book before going to sleep, and a little research will even reveal what book it was, for the graph indicates the succession of eye movements corresponding to lines and to paragraphs. One member of the mission was required to wear a headset of this sort one night out of three. Some of the results obtained follow.

On the average, and irrespective of the number of hours passed in his berth, the subject slept more and more up to the twenty-first day (from five hours to nine and a half hours), then less and less (from nine and a half hours to six and a half hours) from the twenty-first day to the end of the mission.

Up to the tenth day he went to sleep promptly, on the average in seven to eight minutes, more quickly indeed than on any of the days just preceding our departure. Beginning with the fourteenth day it took much more time to go to sleep, once as much as an hour and ten minutes on the twenty-third day.

On the first three days he reached stage four—deep asleep—sooner than before departure. Beginning on the fifth day it was harder for him to reach this stage; on the seventeenth day it took him seven hours to get there; from then on his sleep improved a little but it nevertheless took more than an hour for him to reach stage four.

He dreamed more during the second half of the voyage than during the first.

During one of the last nights aboard he slept eight hours but passed repeatedly from stage one to stage three and even to stage four and back again; that is, he slept badly, but, the report adds, he was not aware of this and believed he had had a good night's sleep.

Independently of the rigorous measurements made with the electroencephalograph, notes and replies to questionnaires prepared in advance show that five members of the expedition went to sleep easily but that on the twenty-second day they were all having trouble falling asleep, trouble that rapidly diminished and attained a "normal" level toward the end of the mission. The sixth member had persistent difficulty in falling asleep during the first fifteen days but from then on went to sleep more and more easily.

SOCIOMETRY

To what extent do the members of an expedition try to isolate themselves? To get together? To form groups of two? Of three? Do they seek to take their meals alone or together? Is there a marked development in these tendencies between the beginning and the end of a mission?

If all this is hardly of definitive importance for a duration of just one month, it may become of prime importance for teams remaining a year or more in a space ship. Certain indications on this subject were obtained but the findings are still fragmentary. In particular, the analysis of some 65,000 photographs taken by the automatic cameras is still to be completed.

The plan had been for the crew members to take meals in teams of two. For this purpose the foodstuffs had been packaged for two persons, and this arrangement was planned, too, for the system of keeping watch and resting.

By one of the teams most meals were taken together up to the tenth day and from then on most were eaten separately. The second team took most meals separately, but this tendency diminished during the first week, only to increase considerably up to the end of the mission. By the third team, meals were most often shared, but the frequency of meals eaten alone also increased with time. The desire for solitude seemed to grow greater as the voyage proceeded.

ANALYSIS OF REFLEXES

I described in Chapter 26 the apparatus we used for this purpose. To tell the truth, the results were not very striking; of the three most indicative sets of data one shows general progress interrupted from the ninth to the eleventh day by a progressive deterioration, followed by a new improvement ending in a constant level for the last nine days of the dive. Another indicates relative stability during the first eleven days followed by a "catastrophic" deterioration on the twelfth day, a deterioration that was gradually disappearing by the end of the voyage. The third group oscillates regularly around the median. Similar results might be apparently obtained by a team doing any sort of work on land. Reflexes may show modifications due to certain passing preoccupations, but "confinement" does not seem to cause significant changes.

ATTITUDE AND HUMOR

The psychologists also tried to grade our good and bad humor. The good humor for one of the passengers fell from the upper limit of 2 to 1 during the first five days and descended slowly to below 1 by the end of the mission. For one of the others, humor dropped from a little more than 1 at the beginning of the dive to close to 0 on the thirteenth day and climbed slowly up to 0.5 at the end of the voyage. A third passed gradually from 2 to 1.5. The humor of two other members of the crew remained nearly constant in the neighborhood of 1 and 1.5 respectively. There are no data for the sixth member.

Despite these figures, which in my view should be treated with caution, good humor reigned almost without interruption from the first to the last day of the mission. It is interesting, nevertheless, to note that, on the average, for the whole crew the number of meals taken together first diminished in the course of the first days only to increase markedly up to the thirteenth day when this number reached its maximum; it was then that the average humor was at its lowest. Also, in the course of this thirteenth day, the electronic detectors showed that the reflexes were most rapid. Reflexes of insulation? Of self-defense? Looking at the personal notes made by the crew, one finds that out of twenty-three entries on this subject fourteen indicate that the writer felt well and nine that he was "content" or "comfortable."

MAINTENANCE OF THE HABITAT

By maintenance is meant supervision as well as upkeep and repair.

There were fifty-four different kinds of inspection to carry out on board, extending from the reading of voltmeters to inspection of the inlets of the cables and pipes through the hull, including checking the air, the water, the depth, and so on. Twenty-six categories had been provided for in advance; twenty-eight had not been directly foreseen and were concerned in particular with the repair of electrical equipment and procedures for disinfection. In all, the operations reached the number of 1,355 instances—that is, an average of 43 a day, and they absorbed from 14 to 20 percent of the man power available each day. Of these operations 1,312 had been foreseen and quickly became routine; 45 had not, and it was on the proper performance of these that the success of the mission largely depended. Moreover, 96 percent of the unforeseen operations of maintenance were carried out by two members of the crew, who thus contributed specifically to the proper functioning of the *Ben Franklin.* These figures show how important it is in a mission of this sort to have on board a good jack-of-all-trades and a complete tool chest.

Hundreds of pages of tables, of statistics, have already been compiled by NASA and by Grumman on this subject; each operation can be sorted out and classified according to its nature, its importance, the chance of recurrence, its consequences for our mission or for a future mission in the sea or in space. A more complete analysis will emerge from our staff. Here I simply wish to emphasize the care with which this area of activity was studied and the sort of conclusions that can be drawn.

HABITABILITY

The concept of "habitability" embraces precise facts about subjects already outlined (temperature, food supplies) and data of a psychological sort which Grumman and NASA graded by different systems. One of these systems deals with the number of "complaints" entered by the crew in the ship's log and in the questionnaires that had to be filled out each day.

Among the "complaints" one must distinguish between the "solicited complaints" and those made voluntarily. Twelve subjects of complaint appeared during the mission; ten were asked for, two were spontaneous; also, five among the subjects asked for were reinforced by voluntary comments.

The solicited complaints were divided as follows:

NO.	SUBJECT
42	communications with the surface and the shore
25	food
24	the seats in the "wardroom"
23	clothing
20	the table in the "wardroom"
20	the beds
22	the inside temperature
19	the accessibility of needed equipment
15	the hot water (which was not hot enough)
12	the small surface of the "kitchen"

The voluntary complaints were divided as follows:

NO.	SUBJECT
14	communication with the surface
3	food
2	clothing
2	the beds
5	the temperature
5	the arrangement of our program
4	noise

The pattern of development of the solicited complaints is as follows:

Concerning the food: 3 on the eighth day; 4 each on the fifteenth and twenty-fourth days; 5 on the twenty-ninth day.

Concerning the clothing: 2 on the eighth day; 4 on the fifteenth day; 5 each on the twenty-second, twenty-fourth, and twenty-ninth days.

Concerning the beds: 4 on the eighth day; 3 each on the fifteenth, twenty-second, and twenty-fourth days; 4 on the twenty-ninth day.

Concerning available space and privacy: 1 on the eighth day; 4 on the fifteenth day; 1 on the twenty-second day; 2 on the twenty-fourth; 3 on the twenty-ninth day.

In my opinion all these figures are difficult to interpret objectively. The complaints increased slightly on the average with time, with a few exceptions, but since they are essentially replies to a questionnaire, it is perfectly evident that the form of the questions had a strong influence on the replies. Here too I am not trying to solve a problem or even to summarize it but simply to show how the specialists attacked it.

MICROBIOLOGY ON BOARD

Here we are dealing with a very different area, one which can be evaluated with any desired degree of accuracy; the quantity of bacilli, of microbes, their names and their development—all this corresponds to well-established norms; the troubles and dangers are known; there are no problems of psychological or psychiatric interpretations, which are always open to question. NASA is especially interested in these microbiological problems for a perfectly obvious reason: how will a crew of astronauts make out on a voyage in space lasting several months? What is to be done in case of an epidemic? One of the voyages planned for the vicinity of the planet Mars envisages a journey of about four hundred days. If a serious epidemic developed in mid-course, that is, in the neighborhood of Mars, the crew would have to "hold out" for at least six months before returning to earth. An investigation of this subject was in NASA's view worth financing, not only for the Gulf Stream expedition but for the construction of the mesoscaph as well, had Grumman not done so.

Five principal zones of inspection were established: the human flora, the environment, the drinking water, food, and clothing.

THE HUMAN FLORA Samples were taken every three days from seven different parts of the body. Sixteen different species of microorganisms were detected and identified; 943 colonies were cultivated. In proportion to the passage of time a kind of simplification of the flora took place in that the number of different species diminished, falling, for example, from fourteen on the first test (the first day of the dive) to eleven on the twenty-eighth day for the whole body; if only the nose, throat, and ears were considered, the figure drops from nine species on the first day to eight on the twenty-eighth day. As a matter of fact, the evolution was irregular, with frequent jumps from one test to another, but the general tendency nevertheless seems clear.

Another manifest tendency is the progressive diminution of what are called Gram-positive organisms and the almost corresponding increase of Gram-negative organisms, whereas these two categories had evolved almost equally before the mission and resumed an almost equal evolution afterward. It must be added that the germicides used had a different effect on these two categories: germicidal soap, which acts directly on the skin, but has very little or no effect on the nose or throat count. For a much longer operation, the microbiological theoreticians might consider purposely introducing Gram-positive germs into the habitat for the purpose of "maintaining the balance."

The bacteria that turned out to be interesting from a medical point of view are the following:

Staphylococcus beta hemolytic aureus, which infested one of the crew during the whole dive but which was not transmitted permanently to any of the others.

Streptococcus beta hemolytic, which turned up on five members of the crew; the same five suffered from a cold at the beginning of the dive; the sixth man did not catch this cold, but on the other hand had a dense colony of *Streptococcus alpha.*

Bacterium anitratum, which was found in the air on the fifth day of the mission, although supposedly it had developed earlier on the skin of the crew. This organism, which had not been noted on the crew before the mission, became widespread by the fourteenth day although no one suffered appreciable harm.

Pseudomonas, an organism considered dangerous in hospitals and often feared more than the *Staphylococcus aureus.* It had been detected on several participants before the start of the mission and it spread rapidly everywhere in the mesoscaph, in the drinking water, on the "furniture," on the bodies of the men. It caused no detectible harm but it represented a serious potential danger, which could have erupted in the case of an open wound, for example.

Proteus, a pathogenic organism that affects the skin in particular. In our case it was transmitted from one of the men to the area of the shower bath and from there to all other members of the crew.

From these first results, two principal conclusions can be drawn: first, that for thirty days the crew "survived" various microbiological "attacks"; and, second, that the common environment during thirty days did not result in completely homogenizing the microorganisms on the six members of the crew; each maintained a whole series of his own microbiological characteristics.

THE ENVIRONMENT Fifteen particular areas were regularly inspected, such as the floor of the forward hemisphere, the floor of the galley, both walls of the forward hemisphere, the principal tables, and the like.

Before departure the interior of the mesoscaph had been carefully cleaned and even partly disinfected. During the voyage the tables and especially the sink in the galley were disinfected daily; insofar as was possible the whole interior of the living quarters was also cleaned and disinfected thoroughly on the seventh, fourteenth, and twenty-first day of the mission.

The tables and plane surfaces were kept in a satisfactory state; samples showed that on departure there were about 10 colonies per square inch, a figure that fell to only a few colonies during the first fifteen days, then slowly rose again and became stabilized during the last days at again about 10.

The interior walls of the mesoscaph varied much more: 0 on departure, 20 colonies per square inch on the fifth day, 7 to 8 on the eighth day, 70 on the eleventh day, 10 to 12 on the fourteenth, 70 once more on the seventeenth day; practically none on the twenty-first day, but a new increase on the twenty-seventh day.

Obviously the floor, the most exposed area, is where the greatest number of colonies per unit of surface were propagated. On departure there were 12 colonies, then 170 colonies on the seventeenth day, followed by a small reduction, then a new advance to more than 200 colonies per square inch on the twenty-fourth day. At the end of the mission the average had been reduced

—about 110 colonies on the twenty-seventh day.

The bacteria in question were chiefly of the Gram-positive type, but the Gram-negative also increased during the mission. To be noted especially are the *Aero-bacters,* the *Proteus,* and the *Pseudomonas.*

Twelve different species were identified and 250 colonies "cultivated."

Again, two principal conclusions can be drawn. First, in proportion to the progress of the mission the flora of the environment and those of the men became more and more alike. Second, systematic cleaning operations lowered the bacteriological level, but each time this level made a considerable advance soon after disinfection.

DRINKING WATER The situation with regard to the drinking water has already been described (page 366). The cold water, which had been disinfected, from the fifth day on served as an effective culture broth for twelve kinds of bacteria, among them the redoubtable *Pseudomonas* and even some coliform bacilli. Basically the cold water had been treated like that in the lunar module, built by Grumman, but in the lunar module the systems and the material used were markedly different from those used aboard the *Ben Franklin.* Moreover, we did not keep up the proportion of iodine in the water to 7.5 parts per million; this proportion seems to have fallen quickly to 0 or thereabouts. The taste of the water, however, remained disagreeable.

Obviously for a new mission of prolonged duration the problem of water should be reconsidered; sterilization by raising the temperature seems to give the best results. Chlorine or even iodine in sufficient quantities to keep the water perfectly aseptic gives it a disagreeable taste, and when most of the food has to be reconstituted with this water the question of taste obviously takes on a further dimension.

Various other biological tests were carried out on board before, during, and after the dive, as circumstances dictated. Minute observations were made of food and clothing; special precautions had been taken to protect the linen and the clothes before and after use, by placing them in airtight plastic bags containing disinfectants.

On the whole, the results were satisfactory, since only one malady occurred aboard, a cold that was almost collective but that lasted barely forty-eight hours. However, we also learned enough on the subject of the behavior of bacteria in the mesoscaph to realize that a new operation of this sort should be preceded by a new and even more detailed study and greater precautions, particularly if the mission were to last for a longer period.

CONVERSION TABLES
BEAUFORT SCALE
INDEX

CONVERSION TABLES

METRIC SYSTEM

UNIT	ABBREVIATION	EQUIVALENT	APPROXIMATE U.S. EQUIVALENT
		Length	
kilometer	km	1,000 meters	0.62 miles
meter	m	100 centimeters	39.37 inches
centimeter	cm	0.01 meter	0.3937 inches
millimeter	mm	0.001 meter	0.03937 inches
		Volume or Capacity	
liter	1		0.2642 gallons
cubic meter	m^3	1,000 liters	264.2 gallons 1.308 cubic yards
		Weight	
metric ton	MT *or* t	1,000 kilograms	2,205 pounds
kilogram	kg	1,000 grams	2.205 pounds
gram	g		0.035 ounces 0.00215 ounces (troy)
		Pressure	
kilograms per square centimeter	kg/cm^2		14.223 pounds per square inch

MONEY

1 Swiss franc	\$0.232 (average)

TEMPERATURES

	DEGREES	
	CENTIGRADE	FAHRENHEIT
freezing point of water	0	32
boiling point of water	100	212

The ratio of the Fahrenheit degree to the centigrade degree is 5:9. Using t_C and t_F to represent respectively the centigrade value and the Fahrenheit value of the same temperature, their relationship can be indicated as

$$t_F = 9/5t_C + 32^\circ, \text{ or } t_C = 5/9\ (t_F - 32^\circ)$$

BEAUFORT SCALE

System for estimating wind speed, first developed by Admiral Sir Francis Beaufort for the British Navy in 1805. The force of the wind is indicated by numbers.

NUMBER	DESCRIPTION	WIND SPEED KNOTS
0	calm	0–1
1	light air	1–3
2	light breeze	4–6
3	gentle breeze	7–10
4	moderate breeze	11–16
5	fresh breeze	17–21
6	strong breeze	22–27
7	moderate gale	28–33
8	fresh gale	34–40
9	strong gale	41–47
10	whole gale	48–55
11	storm	56–63
12	hurricane	64–71

INDEX

ABOUT THE AUTHOR

Jacques Piccard, who now lives in Lausanne, Switzerland, was born in 1922 in Belgium, the son of Dr. Auguste Piccard, the well-known Swiss physicist and pioneer in both outer and inner space.

Jacques Piccard served with the Free French Army in World War Two and was decorated with the Croix de Guerre. He studied at the University of Geneva, received a diploma in history from the Graduate Institute of International Studies in Geneva, and has received honorary Doctor of Science degrees from the American International College, Springfield, Massachusetts, and Hofstra University, in Hempstead, New York.

With his father, Jacques Piccard designed the bathyscaph *Trieste,* which in 1960 completed a record dive of seven miles into the Marianas Trench in the Pacific

Ocean, manned by Jacques Piccard and U.S. Navy Lieutenant Donald Walsh. An account of this project appears in the book *Seven Miles Down.*

Jacques Piccard received the Distinguished Public Service Award from President Eisenhower, the Theodore Roosevelt Distinguished Service Award, and the Richard Hopper Day Memorial Award of the Academy of Natural Sciences of Philadelphia. He is an honorary member of the Institut Suisse Architects Navals and the Société Helvétique des Sciences Naturelles.